Ger

Compact Broadband and Multiband Antenna Designs for Millimeter-Wave Applications

Gerald R. DeJean

Compact Broadband and Multiband Antenna Designs for Millimeter-Wave Applications

Insight into Antenna Architectural Design, Modeling, and Optimization

VDM Verlag Dr. Müller

Imprint

Bibliographic information by the German National Library: The German National Library lists this publication at the German National Bibliography; detailed bibliographic information is available on the Internet at http://dnb.d-nb.de.

Cover image: www.purestockx.com

Publisher:
VDM Verlag Dr. Müller Aktiengesellschaft & Co. KG , Dudweiler Landstr. 125 a, 66123 Saarbrücken, Germany,
Phone +49 681 9100-698, Fax +49 681 9100-988,
Email: info@vdm-verlag.de

Zugl.: Atlanta, Georgia Institute of Technology, Diss., 2007

Produced in USA and UK by:
Lightning Source Inc., La Vergne, Tennessee, USA
Lightning Source UK Ltd., Milton Keynes, UK

ISBN: 978-3-8364-5677-7

To Gerald and Paula

Dad, you taught me the most important lesson that any young man needs to know and that is how to love. Through love, I have been able to clearly see where my passion lies, which is in the field of service. That is the service of human beings. Through your undying love, I began to realize that it really is not important to please myself. And I must admit, when I please others, I feel a sense of satisfaction that overrides any accolade that I have achieved. I know I'll never be as good as you, but I hope I'll get there someday.

Mom, you gave me life. But you definitely didn't stop there. You have always expected nothing but the best from me, and I truly mean nothing but the best. But I guess this is understandable since you have always given me nothing but the best. Even when you didn't have it to give, you still gave. There has never been a limit to the love you have given me through your works. I love you dearly. You are, by far, the greatest woman that I have ever met in my life.

TABLE OF CONTENTS

LIST OF TABLES

LIST OF FIGURES

LIST OF SYMBOLS AND ABBREVIATIONS

1D	One Dimensional
2D	Two Dimensional
3D	Three Dimensional
3G	Third Generation
4G	Fourth Generation
Ag	Silver
AM	Amplitude Modulated
ANOVA	Analysis of Variance
AUT	Antenna-Under-Test
BGA	Ball Grid Array
CAD	Computer-aided Design
COA	Certificate of Authenticity
CP	Circular Polarization
CPW	Coplanar Waveguide
CTE	Coefficient of Thermal Expansion
Cu	Copper
DECT	Digital European Cordless Telephones
DOE	Design of Experiments
F/B	Front-to-Back
FCC	Federal Communications Commission
FEM	Finite Element Method

FDTD	Finite Difference Time Domain
FM	Frequency Modulated
GSM	Global System of Mobile Communications
GTRI	Georgia Tech Research Institute
HFSS	High Frequency System Simulator
HIPERLAN	High Performance Radio Local Area Network
IC	Integrated Circuit
IF	Intermediate Frequency
ISM	Industrial-Scientific-Medical
LCP	Liquid Crystal Polymer
LMDS	Local Multipoint Distribution Systems
LTCC	Low Temperature Co-fired Ceramic
μBGA	Micro Ball Grid Array
MCM	Multichip Module
MEMS	Micro-Electro-Mechanical Systems
MLO	Multilayer Organic
MMIC	Monolithic Microwave Integrated Circuit
MoM	Method of Moments
PBG	Periodic Bandgap
PCS	Personal Communication Services
PEC	Perfect Electric Conducting
PIFA	Planar Inverted-F Antenna
PKCS	Public Key Cryptosystem

PMC	Perfect Magnetic Conducting
RF	Radio Frequency
RFID	Radio Frequency Identification
RLC	Resistor-Inductor-Capacitor
RSM	Response Surface Methodology
SAR	Specific Absorption Ratio
SAR	Synthetic Aperture Radar
SHS	Soft-and-Hard Surface
SIP	System-in-a-Package
SLL	Sidelobe Level
SMD	Surface Mount Device
SOC	System-on-Chip
SOLT	Short, Open, Load, and Thru
SOP	System-on-Package
SPC	Statistical Process Control
SPA	Shorted Patch Antenna
TE	Transverse Electric
TLM	Transmission Line Matrix
TM	Transverse Magnetic
UHF	Ultra High Frequency
VHF	Very High Frequency
VSWR	Voltage Standing Wave Ratio
WLAN	Wireless Local Area Networks

SUMMARY

The following book presents the design, modeling, and optimization of compact antennas architectures for wireless communications, wireless local area networks (WLAN), automotive radar, remote sensing and microwave and millimeter-wave applications. In recent years, the miniaturization of cell phones and computers has led to a requirement for antennas to be small and lightweight. Antennas, desired to operate in the WLAN or millimeter-wave frequency ranges, often possess physical sizes that are too large for integration with radio frequency (RF) devices. When integrating them into three-dimensional (3D) transceivers, the maintenance of a compact size provides isolation from other devices; hence, surface wave propagation does not affect nearby components of the transceiver such as filters, baluns, and other embedded passives. Therefore, the development of a rigorous design method is necessary for realizing compact and efficient antennas in the wireless community. Furthermore, it is essential that these antennas maintain acceptable performance characteristics, such as impedance bandwidth, low cross-polarization, and high efficiency throughout single or multiple frequency bands and standards.

In this work, various compact and packaging-adaptive antennas have been designed for practical wireless communications systems. In addition, research is presented on showing how compact antennas can be applied to other important applications, such as consumer and security applications. A technique to model antennas with equivalent circuits is addressed, while extending the research of compact antennas to produce antenna arrays is also presented. Upon completion of this work, a designer should come away with some novel ideas that can be implemented to design state-of-the-art compact antennas.

CHAPTER 1

INTRODUCTION

Since the early 1970s, microelectronics has been a driving force behind the production of electronic devices in various industries. Some of these industries include the consumer markets, communications, military applications, and automotives [1]. The integration of large numbers of transistors as well as passive components and mixed-signal active devices into a working system started with the design approach of system-on-chip (SOC) [2]. This method allowed for the development of high-performance integrated circuits (ICs) to be produced at a high volume. Some of the drawbacks from this approach were the high production costs resulting from the use of expensive silicon-based chips, low functional reliability attributed to manufacturing, thermal, and mechanical problems, and the limited capabilities of integrating transistors and active (and passive) components on the same chip [3]. In the early 1980s, the advent of the multichip module (MCM) offered an alternative to the restrictions of integrating components on the same chip by integrating multiple ICs (optical, radio frequency (RF), and digital) on a substrate using ball grid arrays (BGAs). This system approach alleviates issues of crosstalk between ICs by allowing the chips to perform their functions independent of each other. The packaging of ICs in MCMs is typically designed in a one- and two-dimensional (1D and 2D) platform [4]. The system-in-a-package (SIP) approach, introduced approximately a decade ago, allows some similar features to MCMs at a lower cost. The big advantage that SIP enjoys over MCMs is the ability to package chips vertically on different layers, thus allowing the production of a three-dimensional (3D) module. Some have called this technique of system integration "SIP-based MCMs", based on the addition of the third dimension in the module. This vertical placement of the ICs allows for more compact modules to be produced. In today's consumer market and for years to come, the integration approach that has the most promise is system-on-package (SOP) technology [5]. The SIP approach uses discrete, bulky passive and active components in its integration [4,5]. The advantage of SOP over SIP modules is the use of thin film embedded components. This allows for the maximal compactness of full scale integrated modules that can be placed in third-generation (3G) and fourth-generation (4G) cellular phones and other wireless modules. SOP technology consists of a multifunction, multichip package that enables the integration of many system-level functions, such as digital, optical, analog, and micro-electro-mechanical systems (MEMS) [6,7]. Figure 1 illustrates a diagram of a high density, highly integrated wireless front-end SOP implementation in a multilayer packaging environment.

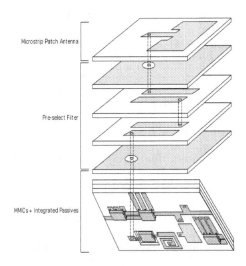

Figure 1. Illustration of high density, highly integrated 3D SOP integration in a multilayer packaging environment.

The ultimate goal of realizing SOP technology is the functional development of such a module. It is seen in this module that the use of vertical integration is critical in the integration of planar devices. The availability of additional dielectric layers allows the three-dimensional deployment of other integrated passives such as baluns, lumped inductors, capacitors, and resistors, as well as intermediate frequency (IF) or low pass filters used in the case of a superheterodyne or a direct conversion scheme, respectively. The integration of the antenna into a compact SOP module continues to be a limiting factor in the development of a complete 3D RF SOP device. Much research has been done in the field of designing antennas that have a large array of functions such as broad impedance bandwidths, multi-frequency operations, and beam steering capabilities. The challenge is to maximize the performance characteristics of antennas while maintaining a compact size.

Due to the emergence of the communications industry, which has taken place over the last 60 years, antennas and its principles has been an important area of focus. Antennas can be incorporated in various geometries and can be used in several applications. Dipole antennas, or "dipoles", (Figure 2a) and monopole antennas, or "monopoles", are common straight wire antennas that exhibit omni-directional radiation. The main difference between dipoles and monopoles is the physical configuration of the antennas. Dipoles have two separate conducting wires, one connected to the positive terminal and the other connected to its negative counterpart. By using image theory, the monopole can be realized by removing one conducting wire and replacing it with a ground plane, hence reducing the size by one half. These antennas can be used to improve reception of frequency modulated (FM) broadcast signals. They are still active

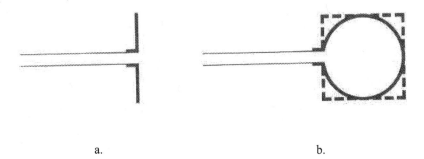

a.   b.

Figure 2. Illustration of (a) a dipole antenna and (b) a loop antenna [8].

in the cellular phone industry as well. Loop antennas (Figure 2b), which were utilized as

far back as the late 1910s, offer very directional radiation. These antennas are a variation

of dipoles and can be constructed by simply shaping a straight wire into a loop. Loop

antennas take many forms, such as rectangular, triangular, and circular, and are primarily

used for amplitude modulated (AM) broadcast and longwave bands. The pyramidal horn

(Figure 3a), conical horn (Figure 3b), and the rectangular waveguide (Figure 3c) are all

examples of aperture antennas that radiate energy through aperture cross-sections. Horn

antennas typically possess a large gain and are used as feeding elements for reflector

antennas. These are also useful for aircraft and spacecraft applications because they can

be flush mounted on the skin of the aircraft or spacecraft [8].

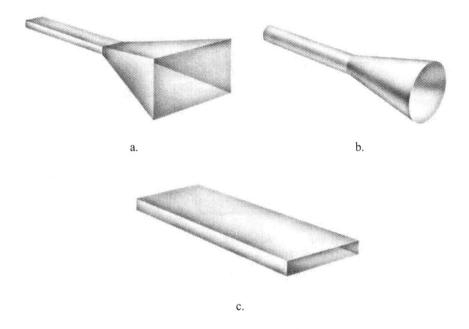

a.　　　　　　　　　　　　　b.

c.

Figure 3.   Illustration of (a) a pyramidal horn, (b) a circular horn, and
(c) a rectangular waveguide [8].

Finally, reflector antennas, such as the parabolic reflector and the corner reflector, allow

the transmission and reception of signals over long distances. These antennas operate by

utilizing a feeding horn antenna that transmits radio waves that reflect off of a large

conducting plane in the opposite direction (Figures 4a and 4b). This is analogous to the

reflection of someone's image as he looks in the mirror. In today's consumer market,

DirecTV is a common service that provides digital television entertainment and

broadband satellite networks to thousands of homes around the world. This is done

through the use of reflector antennas (Figure 5).

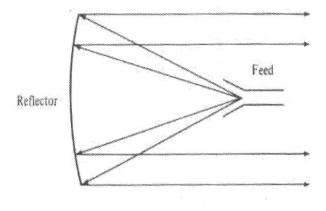

a.

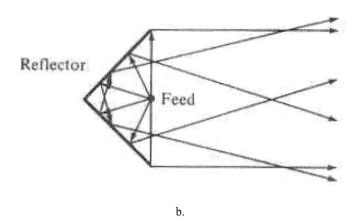

b.

Figure 4. Illustration of (a) a parabolic reflector antenna and (b) a corner reflector antenna [8].

Figure 5. Picture of a DirecTV satellite dish.

Each of the aforementioned antennas has a large physical size that is not suitable for wireless communications transceivers and RF modules. To overcome this limitation, microstrip or planar antennas [8-11] that are printed on a dielectric substrate backed by a conducting ground plane presents a unique solution to designing small structures. An illustration of a microstrip patch antenna is shown in Figure 6. Since these antennas are resonant structures, the need for passive circuitry to assist in achieving resonance is alleviated. Since SOP enables 3D compact architectures to be realized, microstrip antennas have received much consideration for implementation in SOP technology. These antennas can be integrated into 3D modules for millimeter-wave short range broadband communications and reconfigurable sensor networks.

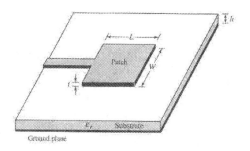

Figure 6. Microstrip antenna [8].

Microstrip antennas enjoy many advantages over their counterparts, such as low manufacturing costs via modern printed circuit technology, a low profile, ease of integration with monolithic microwave integrated circuits (MMICs) and integrated passives, and the ability to be mounted on planar, nonplanar, and rigid exteriors [11].

Low temperature co-fired ceramics (LTCC) multilayer technology is becoming more and more popular for producing complex multilayer modules and antennas because of its flexibility in realizing a variable number of laminated layers [12,13]. Major advantages of using LTCC are lower dielectric loss, size reduction due to a high dielectric constant, hermeticity, the ability to integrate surface mount devices (SMDs) and ICs, and the ability to incorporate embedded passives and interconnect circuitry to be sandwiched between the substrate layers. The high dielectric constant of LTCC is significant in realizing more compact 3D architectures due to the inversely proportional relationship between the dimensions and the dielectric constant. Designing microstrip antennas on LTCC layers offers a desirable approach for integration with RF devices. To increase the physical area of the antenna, one can place air vias in the structure that, will in turn, lower the effective dielectric constant and produce a larger physical area. Some

8

disadvantages associated with LTCC are the shrinkage of the material and the surface wave excitation due to the high dielectric constant of the substrate. Alternative organic materials, such as the liquid crystal polymer (LCP), offer certain advantages over LTCC. These include a lower cost, an engineered transverse coefficient of thermal expansion (CTE), and flexibility, although this is a less mature fabrication technology [14]. In addition, the low water absorption of LCP makes the material stable in a variety of environmental conditions, hence, preventing changes in the material's dielectric constant and loss tangent. The multilayer circuitry can be easily realized due to the two types of LCP substrates that have different melting temperatures. The high melting temperature LCP (around 315 °C) is primarily used as the core substrate layer, while the low melting temperature LCP (around 290 °C) is used as a bonding layer. Therefore, vertical integration can be achieved similarly to LTCC.

When designing planar antennas for wireless communications, it becomes necessary to have a microstrip antenna configuration that is compact in size and able to be integrated with other devices. With the physical area of the antenna being inversely proportional to the frequency, it is sometimes difficult to achieve a compact size for wireless local area network (WLAN) applications that require acceptable efficiency and isolation values. Maintaining a compact antenna that can be optimized for impedance bandwidth, radiation pattern characteristics, high gain, and low front-to-back (F/B) ratios can be utilized in SOP applications as well as RF certificates of authenticity (COAs) for consumer applications. There is often a tradeoff in realizing compact antennas while maintaining performance characteristics.

The focus of this book is the full scale design of compact antenna structures for modern commercial RF systems such as cellular phones for PCS applications, Bluetooth and 2.4 GHz industrial, scientific, and medical (ISM) applications, radio frequency identification (RFID) systems, WLAN (802.11a,b,g,n), LMDS, remote sensing applications, and millimeter-wave applications at 60 GHz. These compact designs will have many performance characteristics such as a broad bandwidth, multi-frequency operation, surface-wave suppressing features, and high gain, "quasi" endfire radiation properties. The performance characteristics of these antennas may vary based on the design application.

Chapters 3, 4, and 5 focus on three important necessities for successful 3D integration of antennas using SOP technologies: compactness, optimized bandwidth, and the improvement of radiation patterns. The first of these chapters presents a design that addresses the need to optimize the impedance bandwidth of integrated antennas using a mature multilayer packaging technology through a simple one parameter technique. Then, a folded shorted patch antenna (SPA) is presented that describes how the resonant length (and therefore, the total size) of a conventional patch antenna can be significantly reduced, thus leading to a miniaturized antenna for any specified frequency. Last, but not least, the third chapter in this section explains a methodology that can lead to the improvement of radiation patterns through the suppression of surface wave modes (that arise when antennas are designed on high dielectric constant substrates) and edge diffraction effects (that arise from placing antennas on large finite size ground planes).

In the next three chapters, the author slightly moves away from the theory of designing compact antennas and discusses a modeling technique that can be integrated to antenna design research as well as some applications where compact antennas can be utilized in arrays. Chapter 6 details how the scattering parameters of antennas are approximated using a rational function approximation developed through a vector fitting approach. This approximation leads directly to the development of equivalent circuits that can assist researchers by providing them with additional information to explain electromagnetic phenomena (such as higher-order modes and parasitics). The next two chapters (7 and 8) discuss the design of high gain antennas through the use of planar arrays. The first of these two chapters presents the characterization of planar arrays at two frequencies (one in the Ku band, while the other above 30 GHz) with dual polarization characteristics designed on LCP multilayer technology. The complexity in this design lies in the configuration and the placement of the arrays on different layers. Factors, such as feedline radiation and blockage, are addressed. The second of these two chapters presents an improved design of a microstrip Yagi antenna array that can achieve a high gain while maintaining a high F/B ratio.

Finally, an initial investigation is presented on the concept and technology of RF COAs that can be used in consumer applications, such as tamper evident seals and physical credit card transactions. This technology presents a unique opportunity to combine the use of electromagnetics, compact antenna design, and cryptography to produce a product that has the potential of creating a long term security solution to a common problem that exists throughout the world, counterfeiting.

This book is comprised of theoretical investigations of the electromagnetic phenomena, equivalent circuit modeling techniques of the scattering parameters, and simulations and measurements of the fabricated prototypes. The compact antenna configurations that are considered in this book consist of patch designs, but other types of microstrip antennas may also be realized. In addition, some of the antenna architectures that utilize a stacked patch configuration or metalized vias shorted to ground are designed on LTCC and LCP multilayer laminates to show the potential of vertical integration in the realization of compact modules.

CHAPTER 2

BACKGROUND

Compact microstrip antenna design has been a "hot" topic of discussion in order to meet the miniaturization requirements of portable communications equipment [15]. Many attempts have been made to decrease the size of antennas, from reducing the substrate thickness to using a substrate with a high permittivity. An overview of previous design methods for compactness and operational maintainability is presented. This is done in three parts. The first part focuses on what has been done to realize compact microstrip antennas. Next, past techniques are described to show how compact antennas can be applied to achieve specific performance characteristics, such as enhanced bandwidth and gain, dual-frequency operation, and circularly polarized radiation. The final part details some complexities and modeling issues associated with designing compact antennas for integrated transceivers and wireless applications.

2.1 History of Compact Design

Various techniques have been documented to reduce the size of microstrip antennas at a given frequency for a compact design. The simplest method is to use a high dielectric constant substrate [8,9,15]. This can be justified by understanding that microstrip antennas are approximately half-wavelength structures, meaning that the resonant length is half of a guided wavelength ($\lambda_g/2$). One guided wavelength can be expressed as the ratio of the phase velocity (v_p) to the frequency (f). Additionally, the phase velocity can be expressed as the ratio of the speed of light ($c \approx 3*10^8$) to the square root of the effective dielectric constant. In equation form, the length of a microstrip antenna is approximately

$$L = \frac{c}{2f\sqrt{\varepsilon_{\text{eff}}}}$$ (1)

The effective dielectric constant increases with increasing dielectric constant [11]. Depending on the process and application, the dielectric constant may remain a fixed parameter. A high dielectric constant will maintain the compactness of the structure, while a low dielectric constant will result in a more efficient radiator. If this dielectric constant has a high value, surface modes may be launched at the interface of the air and dielectric material. Surface waves are transverse magnetic (TM) and transverse electric (TE) modes that propagate into the substrate outside the microstrip patch [16]. These modes have a cutoff frequency that is different than the resonant frequency for the dominant mode of the antenna. Surface waves become a problem when their cutoff frequency is lower than the resonant frequency of the antenna, causing overmoding (more than one propagating mode at a given frequency). The cutoff frequency of a surface wave is inversely proportional to the dielectric constant of the substrate. This is shown in the formula below:

$$f_c = \frac{nc}{4h\sqrt{\varepsilon_r - 1}}$$ (2)

where c is the speed of light (c≈3*10^8), h is the substrate thickness, ε_r is the dielectric constant, and n=1,3,5…for TE$_n$ modes and n=2,4,6…for TM$_n$ modes. If surface waves are present, the total efficiency of the antenna will be reduced. There is often a tradeoff between compact size and efficiency. Therefore, other methods have been proposed to reduce the antenna dimensions with fixed substrate properties. One method is the use of a meandered patch (Figure 7). The meandering is done by cutting slots in the non-radiating edges of the patch [17,18]. This effectively elongates the surface current path on the patch and increases the loading which results in a decrease in the resonant frequency.

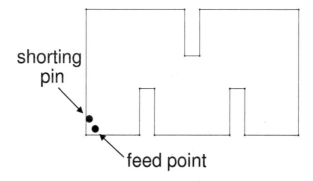

Figure 7. Meandered patch by inserting slots in the non-radiating edges [18].

The tradeoff that is seen in using this method is a decrease in impedance bandwidth and antenna gain, that causes a severe limitation in practical applications [19]. Additionally, high levels of cross-polarization may arise from sections of the meandered patch [20].

Another method includes the meandering of the ground plane [19]. In a similar approach, the insertion of slots in the ground plane can reduce the resonant frequency for a given length (Figure 8).

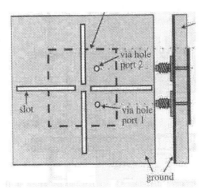

Figure 8.Meandered patch by inserting slots in the ground plane [19].

The slots of the ground plane may cause unwanted levels of backside radiation [21], leading potentially to the high absorption of energy from the human head when the antenna is used in PCS applications, specifically when the antenna is placed in cellular phones [22]. This absorption of radiation to the head is characterized by the specific absorption ratio (SAR). An acceptable SAR level required by the Federal Communications Commission (FCC) for public exposure is 1.6 W/kg [23]. Another popular technique involves a shorted plane that is placed along the middle of the patch parallel to the radiating edge between the patch and the ground plane. With the presence of the shorted plane, half of the patch can be omitted. The patch now has a resonant length of a quarter-wavelength (λ/4). Theoretically, the position of the shorted plane is selected where the electric field normal to the patch is non-existent. Therefore, the fields parallel to the shorted plane are undisturbed. The major disadvantage of this method is a narrower impedance bandwidth that is unsuitable for some applications, such as DECT (digital European cordless telephones) [24]. Also, punching vias through the substrate to create the shorted patch may not be suitable for some materials, such as LCP, because of the alignment inaccuracies of the vias. Targonski and Waterhouse attempted to alleviate this problem by using a thick foam substrate with a low dielectric constant [25], but this affected the compactness of the antenna. Use of the inexpensive foam may not be suitable for RF packaging applications that involve a high temperature environment. Finally, the use of varactor diodes (Figure 9) has been shown to contribute to compact operation by means of tuning the resonant frequency [26]. Varactor diodes contain a capacitance that can be adjusted by changing their voltage. This additional capacitance helps to decrease the resonant frequency, making for a more compact geometry. This structure could also support circularly polarized radiation.

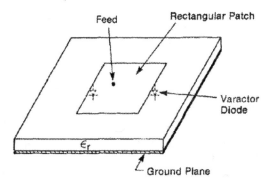

Figure 9. Patch antenna loaded with varactor diodes [26].

One concern of this approach is that the close distance between a varactor diode and the coaxial probe can cause unwanted coupling, while widening this distance by moving the probe may destroy the impedance match. Additionally, although a large bandwidth can be achieved, reduced efficiency and increased levels of cross-polarization are present. Ultimately, the use of varactor diodes presents a problem in terms of integrating the antenna into an RF module. Many times, devices operating in the very high frequency (VHF) and the ultra-high frequency (UHF) bands require antennas to be completely passive elements. Therefore, a varactor diode (an active component) would have to be realized in terms of a printed component instead. Du Plessis and Cloete proposed a solution of using a metallic pad at the radiating edges of a rectangular patch [27]. This design, shown in Figure 10, is completely passive and has the similar feature of changing the resonant frequency of the antenna. When the size of the pads is determined and the antenna is fabricated, a trimming device could be used to trim off metal from the pads.

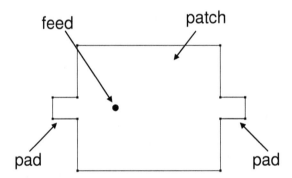

Figure 10. Patch antenna loaded with metallic pads.

By doing this, the antenna's resonant length is shorter because of the shorter surface current path; hence, the resonant frequency can be increased, but by using this method, the frequency of operation cannot be decreased. Also, the trimming of the antenna may affect the performance of the design. The idea of modifying a structure once it has been fabricated is often not practiced in RF packaging and antenna design.

2.2 Previous Techniques in Performance Enhancement

Compact antenna design can be utilized to achieve circular polarization (CP), enhanced gain, and wideband operation for many applications, such as GPS, Bluetooth and WLAN applications. Recent attempts have been made to realize these performance characteristics. One of the simplest ways to attain circular polarization is to insert a cross-shaped slot in the patch [28,29]. This tends to excite two orthogonal modes with a 90° phase difference between them, a necessary condition for CP. This method is useful because it only requires a single feed point. Trimming off the corners of the patch along the same diagonal direction (Figure 11) is another means of achieving CP while maintaining a compact design [30].

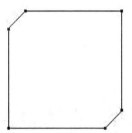

Figure 11. Patch antenna with trimmed corners.

An additional simple approach can be taken where one can slightly increase the length (or width but not both) of a square patch. This makes the patch "nearly square". Using this technique, as well as exciting the patch along the diagonal, achieves CP by obtaining two modes with slightly different resonant frequencies. One mode can "lead" by 45°, while the other mode can "lag" by 45°; hence, a 90° phase difference is produced while maintaining electric field amplitudes that are equal [31,32]. An advantage to this design is the ease in which CP can be achieved. The circuit modeling and radiation characteristics for this approach remain unchanged. Despite the advantages, there has been no formulated approach for choosing the correct length perturbation to achieve CP. There are few designs that have been reported for producing enhanced gain while maintaining compact operation. One of these designs incorporates two substrate layers with the patch antenna embedded between them (Figure 12) [33,34]. The lower substrate (between the patch and the ground plane) has a low dielectric constant ($\varepsilon_r < 5$), while the other substrate layer (above the patch) has a high dielectric constant ($\varepsilon_r > 15$) [34]. This high dielectric constant will excite substrate modes, thus lowering the efficiency and bandwidth which may not be suitable for a desired application [21].

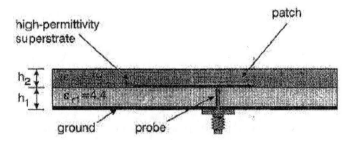

Figure 12. Patch antenna integrated on a low and high permittivity substrate [34].

Another technique that can be used for enhancing the gain of microstrip antennas involves placing parasitic elements next to the radiating patch [35]. The radiating patch will capacitively couple energy to the nearby parasitic elements, creating a wider aperture. Although this wider aperture will increase the gain of the structure, the effect may not be significant. A possible drawback using this approach is the increased lateral area of the design, which may prohibit the compactness of the structure. Careful placement of the parasitic elements must also be taken into account. Placing these elements too close to the radiating patch can greatly decrease the resonant frequency of the antenna, while positioning the elements too far from the radiating patch will exhibit no effects at all. Finally, the achievement of wideband frequency operation has been reported in [15]. In addition to stacking, one design incorporates an aperture-coupled shorted patch with a slot in the ground plane. The uniqueness of this design is the thick air substrate employed under the patch [36]. With the length and width of the patch chosen to resonate at two frequencies that are close to each other, the use of the air substrate helps to widen the bandwidth to a point where it combines to cover the bandwidth of both resonant frequencies. With this design, a total impedance bandwidth of 26% can be achieved. A second compact design with wideband operation utilizes a chip resistor that is placed between the patch and the ground plane at one radiating edge of the structure [37]. The wideband effect can be seen by considering the decrease in the quality factor, or Q-factor, when additional resistance is introduced into the circuit. This decreased Q-factor greatly increases the bandwidth of the antenna, as observed in the equation below:

$$BW = \frac{S-1}{Q\sqrt{S}} \qquad (3)$$

where BW is the bandwidth and S represents the maximum voltage standing wave ratio (VSWR) value that is desired for an acceptable impedance match. For antennas, this value is usually equal to 2. The major disadvantage of this design is the reduced efficiency since a large portion of the input power is dissipated in the resistor, which takes away available power that can be radiated by the antenna.

2.3 Design Complexities

Despite the fact that much work has been done in the area of compact antenna design, there exist some design complexities that must be taken into account. One complexity arises from the use of a high dielectric constant substrate. High dielectric constant substrates are favored for the miniaturization of structures, but surface modes are launched into the substrate which reduces the radiated power, thus significantly reducing the efficiency of the antenna [21]. Another complexity stems from the feeding structure. Some designers prefer to use microstrip lines printed on the same layer as the antenna for excitation. Ease of fabrication and simplicity in circuit modeling are two of the advantages of using these lines. Unfortunately, the radiation loss of microstrip lines increases as the ratio of the square of the length to the square of the free space wavelength (L^2/λ^2) increases [38]. The radiation from the microstrip line may also tilt the main beam a few degrees in the direction of the feed line [39]. A coaxial cable may be suitable for excitation, but depending on the substrate, this may not be possible to manufacture. Additional feeding methods have been utilized when a planar design is necessary. In particular, proximity coupling and aperture coupling are two of the more popular feeding methods because of the decreased levels of cross-polarization and the shielding of the feedline radiation by the ground plane (applicable only in aperture coupling). The proximity-coupled feeding is capacitive in nature; while the aperture-coupled feeding is inductive. The lack of design rules can cause the analysis of both feeding techniques to be complex. Moreover, these feeding methods can only be utilized in a multilayer environment, but they cannot always take full advantage of 3D integration into modules. This places an additional restriction on the design, which may not be suitable for the desired application. It is important to maintain the effectiveness of the antenna by taking into account all complexities that are associated with a design.

2.4 Modeling Complexities

There have been many electromagnetic modeling tools (simulators) introduced in the last ten years for modeling complex structures and high frequency effects of vias and high impedance feedlines. These simulators have been based on many computational algorithms such as method of moments (MoM), finite element method (FEM), and finite difference time-domain (FDTD) schemes. They have been useful in the analysis of the electromagnetic phenomena of passive and active devices. When integrating components together on the SOP module, it is necessary to simulate the electromagnetic characteristics of the entire IC. Because of the physical size of components, as well as the multi-frequency operation of modules, simulating these structures simultaneously may not be feasible for computing systems that have insufficient memory. To obtain a macroscopic insight on the resonance of microstrip patch antennas, an equivalent circuit of a parallel resistor-inductor-capacitor (RLC) network has been proposed in many publications (Figure 13). This circuit representation gives a designer an idea of the resonant frequency and Q-factor of the antenna.

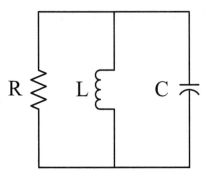

Figure 13. Parallel RLC circuit to simulate a λ/2 microstrip patch antenna.

Design equations for determining the R, L, and C values are presented in [11] based on the physical dimensions of the antenna (length and width of patch and the height of the substrate) and the dielectric constant of the substrate. This circuit is only capable of analyzing single layer microstrip patch designs. For more complex, multilayer antenna structures, additional passive components are needed that can be used to represent the higher order effects of the antenna such as parasitic resonances in the scattering parameter data. Additionally, a more complex passive circuit could be utilized in conjunction with the electromagnetic analysis to form a complete representation of the design.

CHAPTER 3

COMPACT STACKED PATCH ANTENNAS USING LTCC MULTILAYER TECHNOLOGY

The stacked patch antenna is a common approach for achieving a wider impedance bandwidth. One problem with using a stacked structure is the distance between the two antennas may possibly shift the design frequency. In addition, there are many parameters that need to be adjusted for an optimal bandwidth performance, such as the length and width of each patch, the thickness of the substrate, and the position of the feed point. With so many parameters that need to be accounted for, to date, there has been no control over adjusting all variables simultaneously to achieve optimal bandwidth performance. Therefore, a set of design rules is needed to guide antenna engineers in designing wideband antennas.

3.1 Antenna Structure

The integration concept of 3D modules considered in this chapter is illustrated in Figure 14. A stacked-patch antenna is embedded on the top of an RF front-end module in an LTCC multilayer package. The input of the antenna comes from the output of an embedded band-pass filter that is connected with a block of RF active devices by processes called "flip-chipping" and "wire-bonding". The vertical integration capabilities in the LTCC technology provide the space for the embedded RF block. The LTCC cavity process also provides integration opportunities for RF passive components such as switches and/or off-chip matching networks. The vertical board-to-board transition of two LTCC substrates is implemented using a micro ball grid array (μBGA) ball process.

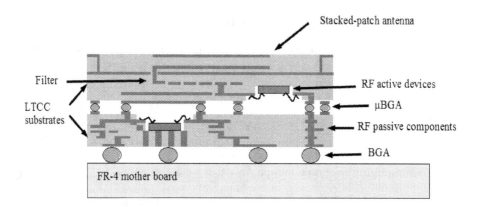

Figure 14. 3D integrated module.

The standard BGA balls insure the interconnection of the highly integrated LTCC module with a mother board such as a FR-4 substrate.

The antenna structure for this design is shown in Figure 15. It consists of two square patches (lower and upper) of length L that are stacked on a grounded LTCC substrate. Square patches were used in this design for the purpose of exciting two orthogonal modes, TM_{10} and TM_{01}, with resonant frequencies that are in close proximity to each other, thereby, obtaining a wider bandwidth. The total thickness of the substrate is h. This thickness can be divided into two smaller thicknesses, h_1, the distance between the lower patch and the ground plane, and h_2, the distance between the lower and upper patch where $h=h_1+h_2$. The lower patch of the antenna structure is excited through a via that is connected to the output port of a filter. A via is a slender piece of metal that vertically connects components on different layers. Then, the electromagnetic coupling of energy is transferred from the lower patch to the upper patch. The position of the via is placed at the center of the radiating edge to match a 50 Ω coaxial line.

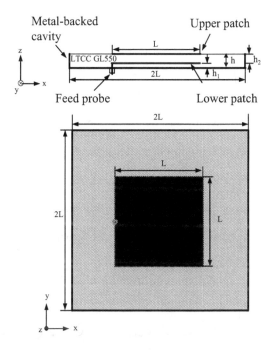

Figure 15. Stacked patch antenna architecture on LTCC multilayer substrate.

In the 3D RF front-end module, the antenna is integrated with other RF circuits. It is essential to prevent any unwanted radiation from other RF components in the integrated module. Therefore, a metal-backed cavity is introduced to shield the RF signals of components surrounding the antenna from the separate antenna signals to preserve functionality. In LTCC packaging technology, a continuous metal wall cannot be realized in fabrication. This obstacle is overcome through the use of an array of vertical vias. It is recommended to choose the lateral dimension of the cavity to be twice that of the stacked patch. A smaller dimension may result in reflections from the walls, which may affect the impedance characteristics of the antenna and hence, the bandwidth of the structure. A larger dimension may hinder the compactness of the structure.

3.2 Theoretical Analysis

Besides the use of square patches in increasing the bandwidth, the major contribution in the wide bandwidth performance of the stacked patch antenna is achieved through the combination of two close resonant frequencies that correspond to the lower patch and the upper patch, respectively. The combination is generated through the electromagnetic coupling between the two patch resonators, which can be modeled by the equivalent circuit shown in Figure 16.

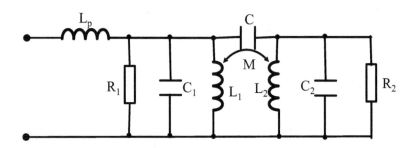

Figure 16. Equivalent circuit of the probe-fed stacked-patch antenna.

This circuit consists of two electromagnetically coupled parallel resonant circuits where L_1 and L_2 are equivalent inductances, C_1 and C_2 are equivalent capacitances, and R_1 and R_2 are radiation resistances. (Subscript 1 refers to the lower patch and subscript 2 refers to the upper patch.) A series inductance, L_p, is included to model the inductance of the feed probe [11]. The two resonant frequencies depend on L_1C_1 and L_2C_2. This is shown in the formula below:

$$f_n = \frac{1}{2\pi(L_nC_n)} \qquad (3)$$

where n = 1,2. Furthermore, the tightness of the electromagnetic coupling is decided by the coupling capacitance C and mutual inductance M. By adjusting the heights of the lower and upper patches, the corresponding resonant frequencies and the coupling tightness can be varied, thus resulting in an optimal impedance performance.

3.3 Preliminary Study

Simulations of this structure using MicroStripes 5.5 were performed, and initial results were taken and analyzed. MicroStripes 5.5 is a 3D fullwave simulator by Flomerics Ltd. that uses transmission line matrix (TLM) modeling for analysis. First, a comparison was done to show the variation of the input impedance as a function of frequency for five values of the relative dielectric constant: ε_r = 2, 4, 6, 8, and 10. This is shown in the form of a Smith chart (Figure 17). A circle labeled "vswr 2:1" represents a reflection coefficient of one-third and a -10 dB return loss. The plot of the input impedance inside this circle shows the -10 dB return loss bandwidth that can be achieved for the particular frequency band. The more input impedance points that lie in the "vswr 2:1" circle, the wider the bandwidth of operation. The horizontal center line of the Smith chart represents a purely resistive impedance. An impedance above this line is resistive and inductive, while an impedance below this line is resistive and capacitive. The impedance loop tends to move downward as the dielectric constant increases. This is a result of the increased parallel plate capacitance resulting from the proportional

relationship between the capacitance and the dielectric constant. Additionally, the

parallel plate capacitance is inversely proportional to the plate separation.

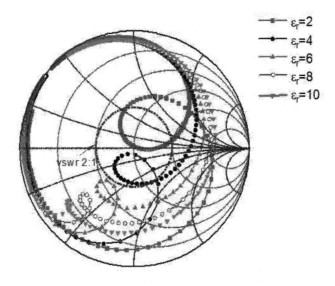

Figure 17. Smith chart of input impedance for variable values of ε_r.

The formula for the capacitance, C, of two finite size plates with a finite separation

distance from each other is shown below:

$$C = \frac{\varepsilon_r \varepsilon_o A}{d} \tag{4}$$

where ε_r is the dielectric constant, ε_o is the permittivity of free space, A is the lateral area

of the plate, and d is the separation distance of the plates. For this simulation, the length and width of both patches are each 10 mm. The distance from the parasitic patch to the

excited patch is 1 mm. The distance from the excited patch to the ground is also 1 mm.

A second simulation was done to analyze the effect of changing the position of

the excited patch. By doing this, the input impedance bandwidth can be optimized.

Figure 18 shows the input impedance as a function of frequency for a fixed dielectric

constant (ε_r = 7) in the form of a Smith chart. It can be observed that a position of 0.5

mm above the ground plane gives an impedance loop that is totally inside the "vswr 2:1"

circle; hence, an optimal bandwidth is achieved. It is also clearly shown that the

impedance loop moves downward as the height of the excited patch (distance from the

ground plane) increases. This simulation was done for a stacked patch structure where

each patch has a length and width of 10 mm. The total substrate thickness is 2 mm.

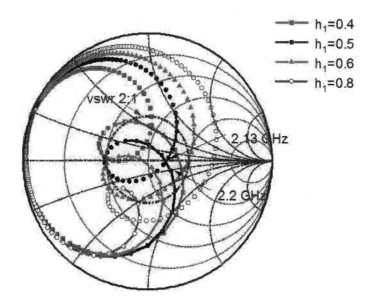

Figure 18. Smith chart of input impedance versus frequency for variable values of lower
patch height at ε_r = 7.

Another simulation was done to also analyze the effect of changing the positioning of the excited patch. Figure 19 shows the input impedance as a function of frequency for a fixed dielectric constant ($\varepsilon_r = 5$) in the form of a Smith chart. It can be observed that a position of 0.6 mm above the ground plane gives an impedance loop that is totally inside the "vswr 2:1" circle, and therefore a bandwidth that is optimal. It is again clearly shown that the impedance loop moves downward as the height of the excited patch above the ground plane increases. This simulation was done for a stacked patch structure where each patch has a length and width of 10 mm. The total substrate thickness is 2 mm.

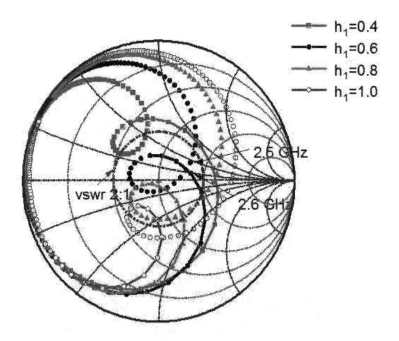

Figure 19. Smith chart of input impedance versus frequency for variable values of lower patch height at $\varepsilon_r = 5$.

A final simulation was done to examine the effect of changing the positioning of the excited patch. Figure 20 shows the input impedance as a function of frequency for a fixed dielectric constant ($\varepsilon_r = 3$) in the form of a Smith chart. The Smith chart shows that a position of 1 mm above the ground plane will give an impedance loop that is totally inside the "vswr 2:1" circle, and therefore an optimal bandwidth. It is again clearly shown that the impedance loop moves downward as the height of the excited patch increases. This simulation was done for a stacked patch structure where each patch has a length and width of 10 mm, and the total substrate thickness is 2 mm.

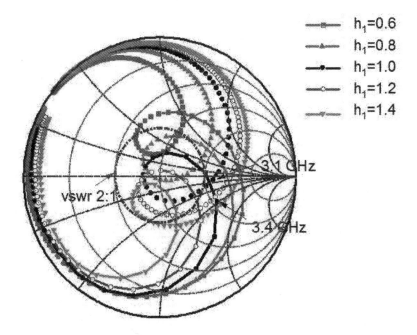

Figure 20. Smith chart of input impedance versus frequency for variable values of lower patch height at $\varepsilon_r = 3$.

3.4 Design Methodology

Upon examination of the simulations and results presented in the last section, a major point of interest can be deduced. When stacked patch antennas were designed on a LTCC Kyocera-GL550 multilayer substrate with a layer thickness of 4 mils per layer, dielectric constant (ε_r) = 5.6, and loss tangent (tan δ) = 0.0012, a relationship between the bandwidth and substrate thickness for a vertically compact structure (substrate thickness = 0.01-0.03 λ_0) is obtained. This is shown in Figure 21, where the relative -10 dB return loss bandwidth (normalized to the resonant frequency, f_r) is plotted as a function of the thickness of the antenna (normalized to the free-space wavelength λ_0 at f_r).

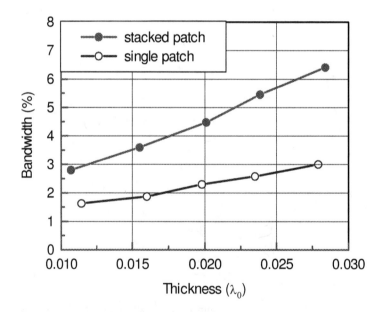

Figure 21. Impedance bandwidth versus total thickness of patch antennas on LTCC multilayer substrate.

The bandwidth of a single patch antenna using the same type of substrate and thickness is presented in this figure as well to show the improvement in the bandwidth when using a stacked configuration. From the plot, it is observed that the compact stacked patch antenna can achieve a bandwidth of up to 7%. This is 60-70% wider than that obtained from a single patch antenna. It is worth noting that the bandwidth of a patch antenna is mainly limited by the dielectric constant and total thickness of the substrate (i.e., the total volume occupied by the antenna). This is understandable by making note of the inversely proportional relationship between the bandwidth and the Q-factor. Moreover, a higher volumetric structure will have a lower Q-factor, and therefore a greater bandwidth.

In lieu of this point and the results obtained in the last section, a set of design rules can be established for the design of stacked patch antennas on LTCC multilayer substrates with ε_r and tan δ close to 5.6 and 0.0012, respectively. The steps are as follows:

Step 1: Choose an initial value for the total design thickness of the substrate, h

($h = h_1 + h_2$). This thickness is usually less 0.05 λ_0 for a compact design.

Step 2: Select the lower substrate thickness, h_1. Through analyzing the results of many simulations using the LTCC Kyocera-GL550 multilayer laminate, it is seen that the impedance loop will be totally inside the vswr 2:1 circle when plotted on a Smith chart, if $h_1 \approx h/4$, which is optimal for an enhanced bandwidth.

Step 3: Design the length L (which is also the width) according to the appropriate resonant frequency, f_r, required for the application. The equation below is suggested for designing the length L:

$$L = \frac{c}{2f_r\sqrt{\varepsilon_r}} \tag{5}$$

where all variables have been previously defined.

Step 4: Determine the upper substrate thickness, h_2, for an optimal return loss. The initial

value of h_2 can be chosen as $3h_1$ according to Step 2. The final value of h_2 may be

obtained by simulation. Upon observation, it is found that the impedance loop in

the Smith chart will move from the upper (inductive) portion of the Smith chart to

the lower (capacitive) portion as the distance between the upper patch and the

lower patch is shortened. The upper substrate thickness, h_2, is determined when

the center of the impedance loop moves closest to the center of the Smith chart,

which corresponds to a minimum return loss and a better matched circuit.

Step 5: Lastly, adjust the length L slightly to cover the desired frequency band. The

simulations will assist in determining the optimal length for the design.

It is possible that the optimized bandwidth of the structure is unsatisfactory for the

desired application. To overcome this dissatisfaction, simply increase the lower substrate

thickness, h_1, and repeat steps 4 and 5 until the required design specifications are met.

3.5 Applications

In the final section of this chapter, the information obtained from initial

simulations and analysis, as well as the design rules that have been postulated, are applied

to three emerging wireless communication bands: the 2.4 GHz ISM band, the IEEE

802.11a 5.8 GHz band, and the 28 GHz LMDS band. The substrate used in these

applications is LTCC Kyocera-GL550 multilayer laminates.

3.5.1 2.4 GHz ISM Band

The 2.4 GHz ISM band has a 3.4% bandwidth with the center frequency, f_c, at 2.4415 GHz. Referring back to Figure 21, a stacked patch antenna with a bandwidth of 3.4% should have an electrical substrate thickness of about 0.015 λ_0. The physical substrate thickness at the specified center frequency is 72 mils. At 4 mils per layer, the total requirement is 18 LTCC layers. By selecting the lower substrate thickness, h_1, to be ¼ the total thickness, four or five layers should be a suitable selection. By using equation 6, the length, L, is 1022 mils. The upper substrate thickness, h_2, can be selected through simulation. The simulated input impedance is plotted for different values of h_2 (in layers) on the Smith chart shown in Figure 22.

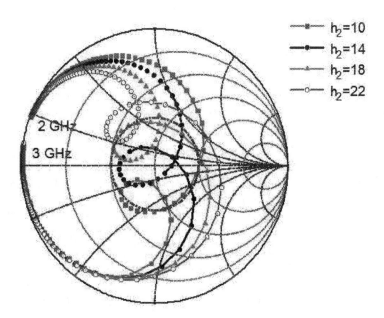

Figure 22. Smith chart of input impedance versus frequency for variable values of upper patch height at 2.4415 GHz.

It can be seen in the Smith chart that the impedance loop tends to move downward in the capacitive region as h_2 decreases. As explained earlier, a higher capacitance is exhibited as the lower and upper patches are closer to each other. When the impedance loop moves closest to the center of the Smith chart and the loop is totally inside the vswr 2:1 circle, a minimum return loss and an optimized bandwidth can be achieved. This is the case for $h_2 = 14$ layers. Therefore, a total substrate thickness, h, of 19 layers (not 18 layers) is necessary for an optimized bandwidth design with h_1 and h_2 equal to 5 and 14 layers, respectively. This coincides closely to the ¼ ratio of h_1 to h. Lastly, the antenna length, L, has to be modified to meet the band specification. Upon simulation, this value is reduced to 966 mils. The input impedance and the return loss versus frequency are plotted in Figure 23. This graph shows the two resonances that are close to each other in the return loss which contributes to a wider bandwidth.

3.5.2 IEEE802.11a 5.8 GHz Band

A similar approach was taken for this application. The selected center frequency is around 5.8 GHz. Referring to Figure 21, the electrical substrate thickness for this band should be approximately $0.015 \lambda_0$. This corresponds to a physical substrate thickness of about eight layers. Initially, h_1 and h_2 are chosen to be two and six layers, respectively. Additionally, the length, L, is set at 400 mils using the formula given in step 3 of the design rules. Once again, the simulated input impedance is plotted for different values of h_2 (in layers) on the Smith chart shown in Figure 24. Coincidentally, the optimized value of h_2 from the Smith chart is six layers. This value agrees perfectly with the ¼ ratio of h_1 to h proposed in step 2.

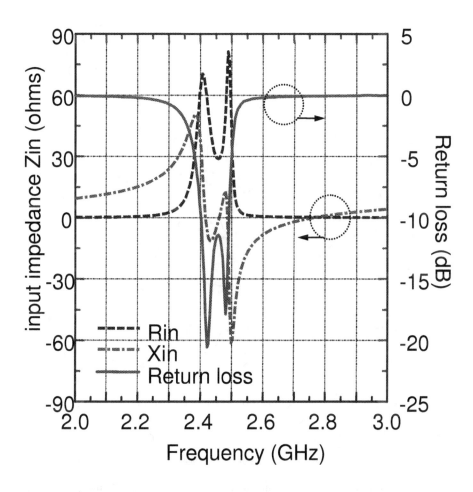

Figure 23. Input impedance and return loss versus frequency of a stacked patch antenna at 2.4415 GHz.

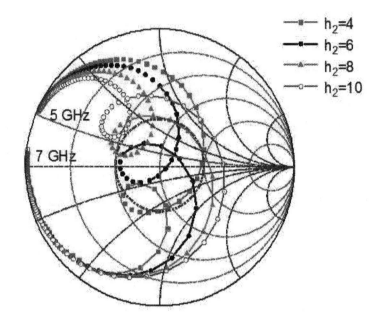

Figure 24. Smith chart of input impedance versus frequency for variable values of upper patch height at 5.8 GHz.

It is worth noting that the ¼ condition can only be met for total layer thicknesses that are multiples of four.

Figure 25 shows the simulated and measured return loss versus frequency of this structure (when h_2 = 6 layers) that was fabricated at Kyocera Industrial Ceramics, Corp. The measured return loss is in good agreement with the simulated results, and a bandwidth of 3.5% is observed. To measure the radiation pattern of the antenna, a modification had to be made in the feeding structure that consisted of a microstrip line connected to the lower patch with a via that passed through the substrate to the top layer and terminated to the surface of a metallic pad. The signal line of an SMA connector was then connected to the pad.

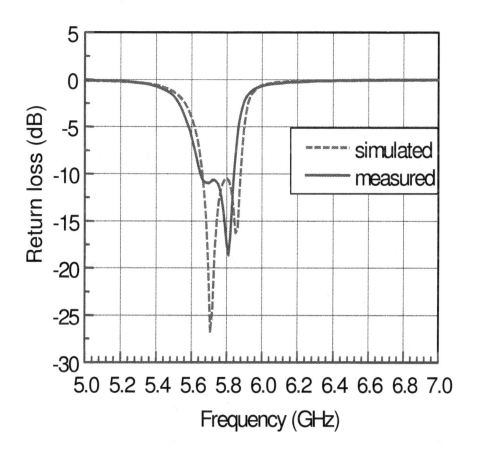

Figure 25. Return loss versus frequency of a stacked patch antenna at 5.8 GHz.

The simulated and measured radiation patterns taken at 5.8 GHz are illustrated in Figure 26. The backside radiation level is about 10 dB lower than the maximum gain, which is about 4.5 dBi. The low gain of this antenna is due to the high dielectric constant and the thin substrate thickness ($0.015\ \lambda_0$). The simulated cross-polarization was less than -40 dB in the E- and H-planes, while the measured cross-polarization was less than -20 dB, which is acceptable for this design. The modified feeding structure caused degradation in the cross-polarization performance of the antenna.

3.5.3 28 GHz LMDS Band

Once again, a similar approach was applied to the LMDS band. The required bandwidth is 7%, and the center frequency is 28 GHz. Based on Figure 21, the electrical thickness is about $0.03\ \lambda_0$ which corresponds to a physical thickness of only three LTCC layers because of the high operational frequency. The height, h_1, is chosen to be a single layer. Due to the low number of layers that is necessary to achieve an optimal bandwidth at this frequency, the ¼ ratio rule of h_1 to h starts to break down. The simulated input impedance is plotted on the Smith chart, shown in Figure 27, for various values of h_2. Since this structure has a thin substrate and a fixed layer thickness, a more extreme variation will exist in the movement of the impedance loops as the value of h_2 changes. As is predicted in Figure 21, this value should be set at two layers to obtain a minimum return loss and an optimized bandwidth. The length, L, of the patch is tuned to 80 mils to fully cover the required band.

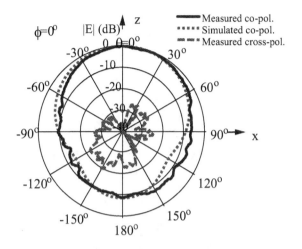

E-plane

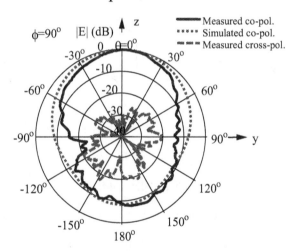

H-plane

Figure 26. Simulated and measured radiation patterns of a stacked patch antenna at 5.8 GHz.

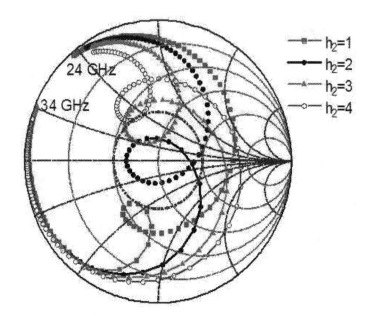

Figure 27. Smith chart of input impedance versus frequency for variable values of upper patch height at 28 GHz.

The input impedance and return loss versus frequency, simulated in MicroStripes 5.5, is plotted in Figure 28 when $h_2 = 2$ layers. For comparison, the return loss versus frequency, simulated in an "in-house" finite difference time domain (FDTD) code, is also shown in Figure 28. From the plots, the two simulators are in good agreement with each other. The return loss is below -15 dB for both plots. The bandwidth for both simulations is close to 7%. The radiation patterns for the stacked patch structure as well as a single patch with the same total thickness, h, is illustrated in Figure 29. The co-polarized (co-pol.) components of the two designs show similar performance for the E- and H-planes. The stacked patch design has a much lower cross-polarization than the single patch design. This is mainly because the single patch has a feed probe that is four

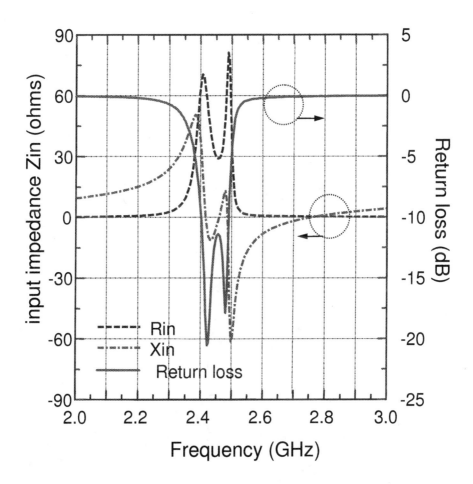

Figure 28. Input impedance and return loss versus frequency of a stacked patch antenna at 28 GHz.

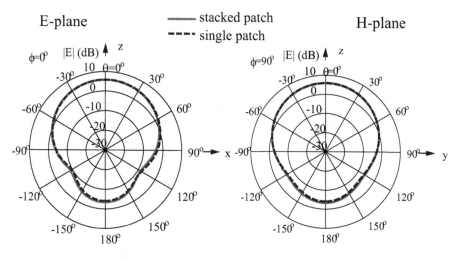

(a) co-polarized component

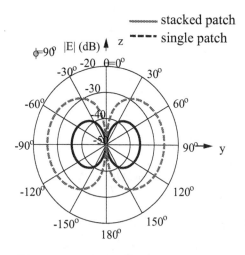

(b) cross-polarized component

Figure 29. Radiation pattern performance comparison of a stacked patch versus a single patch antenna at 28 GHz.

times longer than the stacked patch structure, therefore contributing to higher cross-polarization. All of the designs presented have a radiation efficiency greater than 85%. The power lost in these structures is due to conductor loss (conductivity, σ, equal to $5.8*10^7$ siemens per meter, S/m), dielectric loss (tan $\delta = 0.0012$), surface wave loss ($\varepsilon_r = 5.6$), and feedline radiation. The first three types of loss are properties of the metal (copper, Cu) and the substrate.

Designing thin feeding structures and transitions can circumvent the problem of feedline radiation. It is worth noting that the size of a single patch design would have to be doubled to achieve the same bandwidth as the stacked patch antenna. Therefore, the stacked patch antenna is a great solution for the vertical integration of wireless transceivers using multilayer substrates such as LTCC, LCP, and multilayer organic (MLO).

CHAPTER 4

FOLDED SHORTED PATCH ANTENNAS

The planar inverted-F antenna (PIFA) is one of the most well known and documented small patch antennas. Actually, the PIFA can be viewed as a shorted patch antenna. Therefore, the antenna length of a PIFA is generally less than $\lambda_0/4$. When the shorting post is located at a corner of a square plate, the length of the PIFA can be reduced to $\lambda_0/8$. The size of a PIFA can be also reduced to less than $\lambda_0/8$ by capacitively loading the antenna.

Recent research efforts on the size reduction of patch antennas have focused on the patch shape optimization to increase the effective electric length of the patch. A three-layer or two-layer, folded rectangular patch may reduce the resonant frequency by 50% compared to a conventional patch. Additionally, a printed antenna with a surface area 75% smaller than a conventional microstrip patch can be obtained by incorporating strategically positioned notches near a shorting pin.

To optimize the level of compactness in microstrip antenna design, it is necessary to develop a design that has two properties. The first has a reduced resonant frequency that is a direct result of the reduced resonant antenna length. The second has a reduced resonant frequency that is independent of the total size of the antenna. To achieve both of these properties, simulations and analysis are done to examine the sensitivities in the design with respect to tuning the resonant of the design. Additionally, radiation pattern performance must be maintained. This chapter examines a simple technique for reducing the overall size of a conventional microstrip antenna without the use of external loads.

4.1 Antenna Structure

The antenna structure considered in this chapter is shown in Figure 30. It consists of two square patches that are stacked for bandwidth enhancement. The dimensions of the lower patch are denoted by length, L_1 and width, W_1. Similarly, the upper patch is denoted by L_2 and W_2 for the length and width, respectively. These antennas are supported by two metal shorting walls, one for each patch that are parallel to each other (with respect to the x-axis shown in Figure 30) on different radiating edges. The distance between the ground plane and the lower patch is h_1, while the distance between the lower patch and the upper patch is h_2. The total height of the antenna (denoted by h) is the sum of these two smaller dimensions ($h_1+h_2 = h$). There is no substrate for this design; this is favorable for designing highly efficient radiators with a maximal bandwidth.

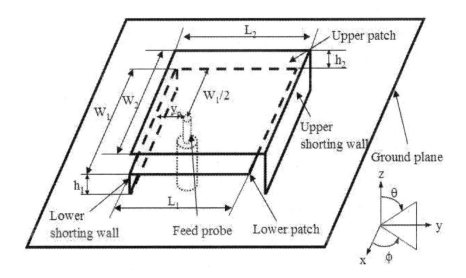

Figure 30. Antenna structure of the folded SPA.

Furthermore, placing this structure next to other components should not affect the functionality of those components due to the inexistence of surface waves in free space. Hence, a cavity is not necessary for this design. The antenna is excited with a 50 Ω coaxial probe in which the probe's signal pin is connected to the lower patch. The charges from the lower patch electromagnetically couples energy to the upper patch which, in turn, radiates space waves. The probe is placed along the center line perpendicular to the lower patch width, W_1, and along a line parallel to the lower patch length, L_1, in order to match the 50 Ω feed probe. The distance between the radiating edge of the lower patch and the feed probe is y_p. This parameter is varied to find the best match between the probe and the antenna. The dimensions of the ground plane should be, at least, twice the size of patches but no more than four patch sizes to maintain compactness and prevent edge diffraction effects from degrading the radiation pattern.

4.2 Design Methodology

The development of this structure comes from the simple design of a patch antenna. A conventional rectangular patch antenna operating at the fundamental mode (TM_{10} mode) has an electrical length of $\sim\lambda_0/2$. This is illustrated in Figure 31a. If one considers that the electric field for the TM_{10} mode at the middle of the patch is zero, the patch can be shorted along its middle line with a metal shorting wall without significantly changing the resonant frequency of the antenna. By doing this, the resonant length of the antenna becomes $\sim\lambda_0/4$ (shown in Figure 31b) and half of the patch can be removed. The physical length is now $L_1/2$. Then, the side of the antenna opposite the shorting wall is folded along the middle of the patch. Simultaneously, the ground plane is also folded along a position that is a short distance from the middle of the patch (Figure 31c). After this procedure, the resonant length of the antenna remains $\sim\lambda_0/4$, while the physical length is reduced by a half (to $\sim\lambda_0/8$). The folding of the ground plane produces the upper patch and its shorting wall. It should be emphasized that it is necessary to fold the ground plane while folding the shorted patch. Otherwise, the folded antenna would look like an S-antenna developed in [40] for a dual-band operation. As a result of this step, the length, L_1, will be slight smaller than L_2 due to the small gap between the lower patch and upper patch shorting wall. Finally, a new piece of ground plane is added to the right of the existing ground plane to regenerate its original length. The new lower patch is created by pressing the two portions together. The completed structure is shown in Figure 31d. The lateral size of a conventional patch antenna has been effectively reduced from a $\sim\lambda_0/2$ to a $\sim\lambda_0/8$ structure. This is a major contribution to the goal of producing compact antennas for commercial use.

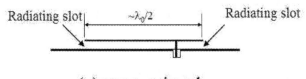

(a) conventional
rectangular patch

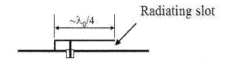

(b) conventional SPA

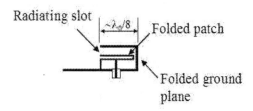

(c) folding of a
conventional SPA

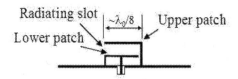

(d) folded SPA

Figure 31. Development of a folded SPA.

4.3 Design Validation

To demonstrate the potential of the design methodology proposed in the last section, two antennas were simulated using MicroStripes 5.6 and an "in-house" FDTD simulator: a standard SPA (resonant length = $\sim\lambda_0/4$) and the folded SPA proposed here (resonant length = $\sim\lambda_0/8$). The medium is free space. The lateral dimensions of the antennas are 10 mm x 10 mm. The patch of the standard SPA is 0.5 mm above the ground plane. For the folded SPA, the total thickness of the structure is 3 mm, and the lower patch is 0.5 mm above the ground. Figure 32 shows the return loss versus frequency for the two designs.

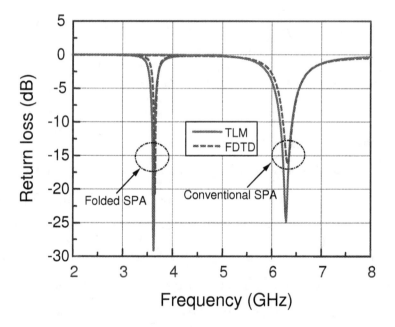

Figure 32. Return loss versus frequency of folded SPA compared with a conventional SPA.

Good agreement is observed between the two simulators. The standard SPA has a resonant frequency of around 6.3 GHz, while the folded SPA resonates at approximately 3.6 GHz. Therefore, the antenna length of the folded SPA is only $\lambda_0/8$; this is about four times smaller than a conventional patch antenna. Additionally, these results display a resonant frequency reduction of about 43%. This is lower than the expected value of 50%. The reason for this discrepancy is due to the slight reduction in the length of the lower patch, L_1, (= 9.5 mm) needed to create the small gap between the lower patch and the shorting wall. This gap allows the power to flow from the feeding point to the radiating slot. To test this explanation, the input impedance of the folded SPA is plotted on the Smith chart illustrated in Figure 33. The lower patch length and width are 9.5mm and 10 mm, respectively. In the same figure, the input impedance of another standard SPA is displayed. The lateral dimensions of this standard SPA are 19 mm x 10 mm. It is worth noting that the length of the standard SPA is twice as long as the folded SPA so the resonant frequency should theoretically be the same. The results from the Smith chart show that the input impedance curves are quite similar with the resonant frequency of the antennas differing by less than 3%.

Next, the near field electric fields and the surface currents for the folded SPA and the standard (19 mm x 10 mm) SPA are investigated and shown in Figure 34. It can be observed that the folded SPA has a field distribution close to that of the standard SPA. In the folded SPA, intense levels of electric fields are concentrated between the lower patch and upper patch, while similar field intensities for the standard SPA are observed opposite the shorting wall. The surface currents exhibit similar performance in examining both structures. The lower and upper surfaces of the lower patch of the folded

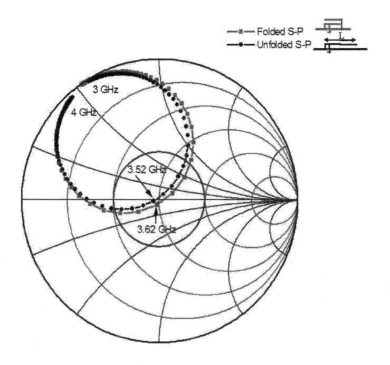

Figure 33. Smith chart of input impedance versus frequency for folded SPA and a standard SPA.

SPA correspond to the lower surface of the standard SPA. Analogously, the lower surface of the upper patch of the folded SPA corresponds to the right-half part of the ground plane beneath the standard SPA. This proves that the path of the electric fields and surface currents of the folded SPA has electrically "folded over" with the physical folding of the antenna. From the enlarged plot of the electric field distribution in the folded SPA, an electric field concentration between the edge of the lower patch and the shorting wall of the upper patch is observed due to the effects of a sharp edge of the lower patch and the short distance between the edge and the shorting wall. This electric field concentration may lead to a reduction of the impedance bandwidth. One way to

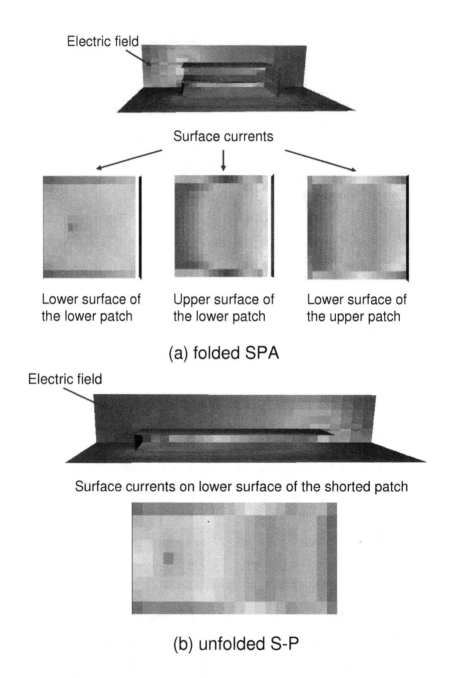

Electric field

Surface currents

Lower surface of
the lower patch

Upper surface of
the lower patch

Lower surface of
the upper patch

(a) folded SPA

Electric field

Surface currents on lower surface of the shorted patch

(b) unfolded S-P

Figure 34. Electric field and surface current distributions at the resonant frequencies.

alleviate this problem would be to increase the distance between the edge and the shorting wall (decreasing length, L_1). However, a decrease in L_1 will result in an increase in the resonant frequency. For this example, the resonant frequency is 3.6 GHz with a -10 dB return loss bandwidth of 1.9% for L_1 = 9.5 mm, while the resonant frequency increases to 3.8 GHz with an increased bandwidth of 2.1% when L_1 is decreased to 8.5 mm. The upper patch length, L_2, remained 10 mm for this example. The difference in the bandwidth of the folded SPA and the standard SPA is approximately 2%.

The E- and H-plane radiation patterns of the folded SPA (f_r = 3.6 GHz) and the standard SPA (f_r = 6.3 GHz) are presented in Figure 35. The radiating slots of both structures are oriented in the same direction. The directivity of the folded SPA (3 dBi) is 1 dBi lower than that of the standard SPA (4 dBi) as seen in the E_φ component. Moreover, the backside radiation level is larger for the folded SPA as well. This is due to the smaller ground plane of the antenna. A ground plane that is between 2λ - 3λ in size will have a significant effect in suppressing the backside radiation. In practice, there is a tradeoff between reducing the backside radiation (using a large ground plane) and designing a compact integrated structure (using a small ground plane). The radiation efficiencies of the standard and folded SPA are 96% and 94%, respectively. The slight reduction is a result of the strong surface current distribution on the folded SPA and the additional conductor loss on the shorting wall of the upper patch. The metal used in the simulations was copper (σ = 5.8×10^7 S/m).

An additional investigation is performed to further reduce the antenna's resonant length from ~$\lambda_0/8$ to ~$\lambda_0/16$. Figure 36 shows the return loss versus frequency for the 10 mm x 10 mm folded SPA when the lower patch is vertically placed at five positions

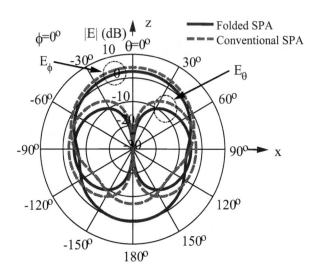

E-plane

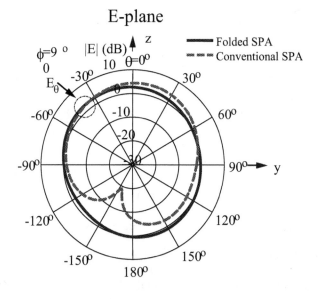

H-plane

Figure 35. Radiation patterns of folded SPA and standard SPA at 3.6 GHz.

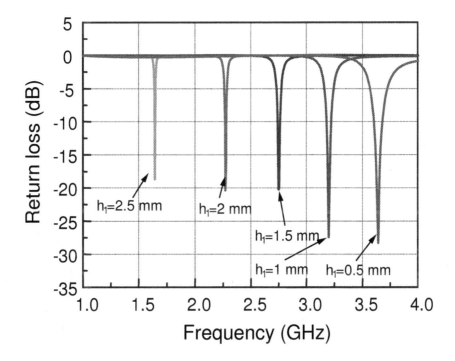

Figure 36. Return loss versus frequency for different vertical placements of lower patch.

between the ground plane and the upper patch. The total thickness remains 3 mm. There is a definite relationship between the lower patch height and the resonant frequency. As the height is increased, the resonant frequency is greatly reduced. By placing the lower patch 2.5 mm above the ground plane, the resonant length of the antenna can be transformed from $\sim\lambda_0/8$ to $\sim\lambda_0/16$. There is a major setback for this approach. As the upper and lower patches become closer to each other, the radiation efficiency is greatly affected. Again, this is due to the strong coupling between the plates. For a folded SPA with an antenna length of $\sim\lambda_0/16$, the radiation efficiency is less than 40%. Careful

attention must be maintained when adjusting this parameter.

4.4 Theoretical Analysis

Some insight on the equations and theoretical justifications that govern the

proposed design are presented in this section. In order to avoid repetition and confusion,

the author has decided to utilize the theoretical analysis presented in [41].

The impedance performance of the folded SPA antenna can be analyzed by

employing a simple transmission-line model. Consider a folded SPA with three different

patch height arrangements: case 1 ($h_1 = h_2 = 1.0$ mm), case 2 ($h_1 = 0.5$ mm, $h_2 = 1.0$ mm),

and case 3 ($h_1 = 1.0$ mm, $h_2 = 0.5$ mm). The equivalent standard SPA configurations

associated with these three cases are illustrated in Figures 37a-37c. By neglecting the

effect of discontinuities because $|h_1 - h_2|$ is much smaller (at least ten times less) than the

length of the folded SPA, the standard SPA can be represented by a transmission-line

equivalent circuit as shown in Figure 37d with input impedance

$$Z_{in} = jX_f + Z_1 \tag{6}$$

where X_f is the feed probe reactance given by

$$X_f = \frac{\omega\mu_0 h_1}{2\pi}\left[\ln\left(\frac{2}{\beta * r_p}\right) - 0.57721\right] \tag{7}$$

with $\beta = 2\pi / \lambda_0$ and the feed probe radius denoted by r_p. Z_1 ($= 1/Y_1$) is obtained from the

transmission-line equivalent circuit, that is,

$$Y_1 = Y_{01} \frac{1}{j\tan(\beta y_p)} + Y_{01} \frac{Y_2 + jY_{01}\tan\left[\beta(L_1 - y_p)\right]}{Y_{01} + jY_2\tan\left[\beta(L_1 - y_p)\right]} \tag{8}$$

and

$$Y_2 = Y_{02} \frac{Y_s + jY_{02}\tan(\beta L_1)}{Y_{02} + jY_s\tan(\beta L_1)} \tag{9}$$

where Y_{01} and Y_{02} are the characteristic admittances of the lower and upper patches,

respectively, and $Y_s = G_s + jB_s$.

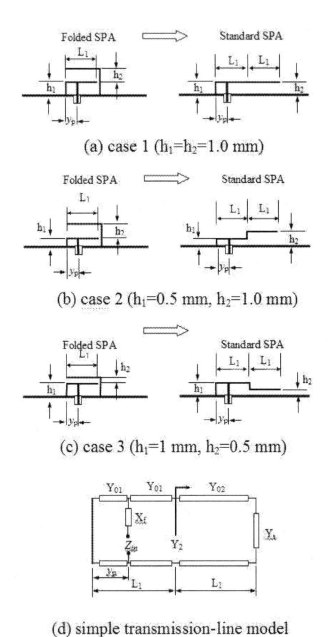

(a) case 1 ($h_1=h_2=1.0$ mm)

(b) case 2 ($h_1=0.5$ mm, $h_2=1.0$ mm)

(c) case 3 ($h_1=1$ mm, $h_2=0.5$ mm)

(d) simple transmission-line model

Figure 37. Folded SPA and its equivalent transmission-line model.

Here, G_s is the conductance associated with the power radiated from the radiating edge (radiating slot), and B_s is the susceptance due to the energy stored in the fringing field near the edge. In the calculations, the following equations were used for Y_0 (= Y_{01} for h = h_1 or Y_{02} for h = h_2), G_s, and B_s

$$Y_0 = \frac{W/h + 1.393 + 0.667\ln(W/h + 1.444)}{120\pi} \qquad \text{for W/h} \geq 1 \qquad (10)$$

$$G_s = \begin{cases} W^2/(90\lambda_0^2) & for & W \leq 0.35\lambda_0 \\ W/(120\lambda_0) - 1/(60\lambda_0^2) & for & 0.35\lambda_0 \leq W \leq 2\lambda_0 \quad (h_2 \leq 0.02\lambda_0) \\ W/(120\lambda_0) & for & 2\lambda_0 \leq W \end{cases} \qquad (11)$$

$$B_s = Y_{02}\tan(\beta\Delta l) \qquad (12)$$

$$\Delta l = \frac{\zeta_1\zeta_3\zeta_5}{\zeta_4}h_2 \qquad (13)$$

where W is the width of the patch and coefficients ζ_1, ζ_3, ζ_4, ζ_5 can be found in Appendix B of [11].

The theoretical results for the input impedance have been obtained using the above analytical expressions and compared in Figure 38 with simulations (using MicroStripes 5.6) for the above three cases of the folded SPA, demonstrating a good agreement. The difference between the theoretical and simulated resonant frequencies is less than 3%. It can be observed that the resonant frequency decreases as h_2/h_1 decreases. For simplicity, one can neglect the effects of Y_s (typically $Y_s \ll Y_0$) and X_f (note that we are now only interested in the resonance of the patch alone). As a result, the standard SPA becomes a shorted transmission-line loaded with an open transmission-line.

63

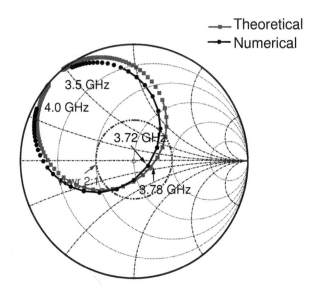

(a) case 1 ($h_1=h_2=1.0$ mm)

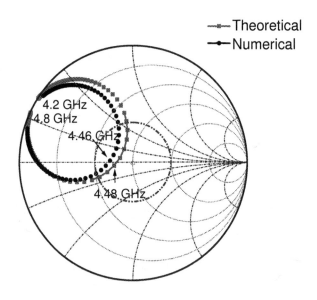

(b) case 2 ($h_1=0.5$ mm, $h_2=1.0$ mm)

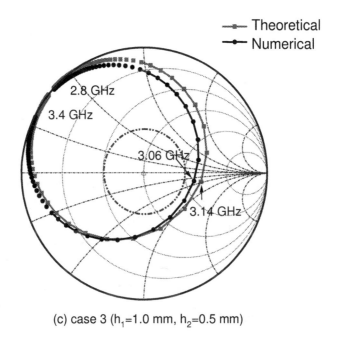

2.8 GHz

3.4 GHz

3.06 GHz

3.14 GHz

(c) case 3 (h_1=1.0 mm, h_2=0.5 mm)

Figure 38. Smith chart of input impedance versus frequency for cases 1-3.

Assuming that the resonant frequency is almost independent of the feeding position, one can choose $y_p = L_1$. Thus, Y_1 becomes

$$Y_1 = Y_{01} \frac{1}{j \tan(\beta L_1)} + jY_{02} \tan(\beta L_1) \tag{14}$$

At resonance, $Y_1 = 0$, and this leads to

$$Y_{01} / \tan(\beta L_1) = Y_{02} \tan(\beta L_1) \text{ or } \tan(\beta L_1) = \sqrt{Y_{01} / Y_{02}} \tag{15}$$

From equation 11, it is observed that Y_0 is inversely proportional to h; therefore, it is observed from equation 16 that the resonant frequency varies proportionally with h_2/h_1.

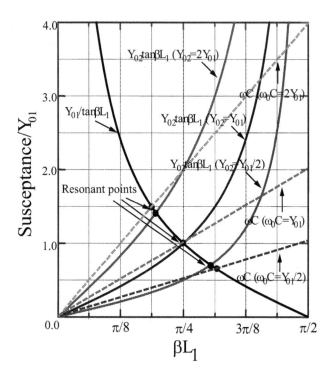

Figure 39. Graphical solution of equations 10 and 11 for the calculation of the resonant frequencies of a capacitively loaded S-P ($\omega_0 = 3\pi/(4L_1) \times 10^8$ rad/sec).

A graphical solution of equation 16 for resonant frequency is depicted in Figure 39, where the intersection of the curves $Y_{01}/\tan(\beta L_1)$ and $Y_{02}/\tan(\beta L_2)$ implies a resonant point. Again, it is observed that if $Y_{01} = Y_{02}$, then $\beta L_1 = \pi/4$, which corresponds to an antenna length of $L_1 = \lambda_0/8$. Also, it is evident that an increase in Y_{02} leads to a decrease in βL_1 if Y_{01} remains unchanged. Considering the upper patch as a capacitive load leads to a clear picture of the physical insight for the above analysis. Replacing the upper patch with a capacitor, C, which is connected between the radiating edge of the lower patch and the ground plane, equation 15 becomes

$$Y_{01} / \tan(\beta L_1) = \omega C .\qquad\qquad (16)$$

A graphical solution of equation 16 is also plotted in Figure 39. It is obvious that the resonant frequency decreases as the capacitance increases. The resonant length of a capacitively loaded SPA will reduce to $L_1 = \lambda_0/8$ if the loaded capacitance is $C = Y_{01}/\omega_0$, where $\omega_0 = 3\pi / (4L_1) \times 10^8$ radians/second, obtained from $\beta L_1 = \pi/4$. Actually, a decrease in h_2 is equivalent to an increase in the coupling capacitance between the upper and lower patches, thus eventually leading to a decrease in the resonant frequency. In fact, some of the small antenna structures researched today can be considered as capacitively loaded patches.

It is noted that the above simple transmission-line model works well when the total height of the folded patch is much smaller (at least five times less) than the patch length and when the discontinuity ($|h_1 - h_2|$) is much shorter (at least 10 times less) than the total length of the folded patch antenna.

4.5 Applications

A practical design of a folded SPA has been constructed and simulated to operate in the 2.4 GHz ISM band. Although the patch dimensions are 15 mm x 15 mm ($\sim\lambda_0/8$ x $\sim\lambda_0/8$), the upper patch is slightly greater in order to create the gap between the lower patch and the upper shorting wall necessary to maintain the path of electric fields. The antenna (fed by coaxial probe) is positioned 5 mm from the lower shorting wall. The total thickness is 6 mm. The lower patch is elevated 2.85 mm above the ground plane. This placement allows the resonant frequency to be tuned to 2.44 GHz. This antenna was fabricated by the Georgia Tech Research Institute (GTRI), where measurements were also taken. The finished prototype is shown in Figure 40. The simulated and measured

return loss versus frequency is illustrated in Figure 41. Good agreement is observed in

the two plots. The slight frequency shift is primarily due to a dimensional inaccuracy in

the fabrication.

Lower patch   Upper patch

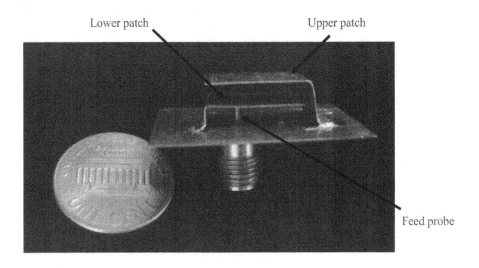

Feed probe

Figure 40. Prototype of folded SPA at 2.4 GHz.

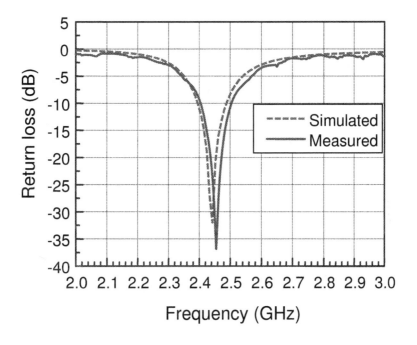

Figure 41. Return loss versus frequency of folded SPA at 2.4 GHz.

The radiation patterns for the E- and H-plane are displayed in Figure 42. Once again, good agreement is seen in the E_θ and E_φ components. The high levels of backside radiation are due to the existence of a small ground plane (30 mm x 30 mm) with respect to the patch dimensions. This can be improved by implementing a compact periodic bandgap (PBG) or a soft-and-hard surface (SHS) structure.

As a point of interest, the folded SPA can be designed using a multilayer substrate such as LTCC or LCP. In order to integrate this antenna into a multilayer package, the shorting walls would be replaced by rows of vias. The via-to-via spacing is a critical parameter to control. Spacing the vias too far from each other will have no significant effect in keeping the electric fields confined under the lower and upper patches. On the other hand, an extremely close spacing will cause some parasitic capacitance and additional currents from the vias that may affect the return loss and reduce the efficiency of the antenna. Additionally, the probe can be replaced by a via that passes through the ground plane and terminates on the top of a microstrip line that is under the ground plane. A complete analysis (through simulation and theory) is necessary to effectively design this structure to be integrated into a compact 3D module.

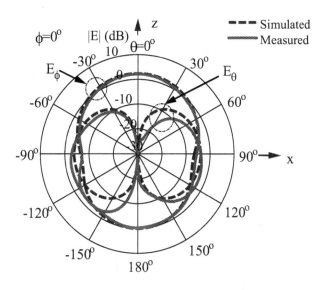

E-plane

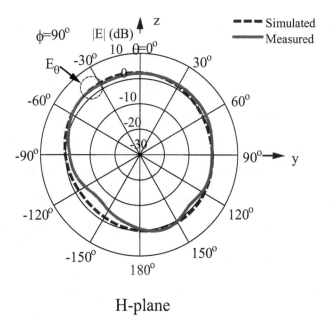

H-plane

Figure 42. Radiation patterns of folded SPA at 2.4 GHz.

CHAPTER 5

SUPPRESSION OF SURFACE WAVES AND RADIATION
PATTERN IMPROVEMENT USING SHS TECHNOLOGY

As multilayer materials such as LTCC, LCP, and MLO are being utilized for the integration of antennas into SOP modules, there are some concerns that need to be addressed in order to maintain the performance of the antenna. One of the drawbacks of designing antennas on LTCC is the high dielectric constant of the substrate ($\varepsilon_r > 5$), which facilitates the propagation of surface waves, which may be a larger problem at higher frequencies (millimeter-wave range). Designing antennas on high dielectric constant substrates can severely degrade the performance of the antenna's radiation characteristics as well as reduce the efficiency of the radiator. Another disadvantage of the surface wave propagation due to antennas on high dielectric constant substrates is the unwanted coupling of energy between the antenna and other active devices on the module. Although placing the antenna at a distance of about three wavelengths (3λ) away from the active devices can reduce the crosstalk between devices, this approach is not feasible for maintaining a compact module. Additionally, the use of vertical integration capabilities associated with SOP technology has allowed the lateral size (length and width) of the substrate to be decreased considerably. Despite this innovation, planar antennas that are designed on small-size substrates contribute to backside radiation below the ground of the structure. There is a need for a surface wave suppressing mechanism that can be integrated to the patch antenna to maximize the performance of the antenna and minimize the degradation of other devices on the module. The most common method of suppressing surface waves is the use of a periodic bandgap structure (PBG). Sometimes, integrating PBG structures into SOP-based devices may not be suitable for maintaining the compactness of the design because PBG structures can be quite large as a result of the rows of via holes needed to realize the bandgaps. A new implementation using the soft surface properties of a soft-and-hard surface (SHS) structure is applied to a patch antenna on LTCC multilayer substrates.

5.1 Theoretical Analysis

The ideal SHS conditions can be characterized by the following symmetric boundary conditions for the electric and magnetic fields [42]:

$$\hat{h} \cdot \bar{H} = 0 \qquad \hat{h} \cdot \bar{E} = 0, \tag{17}$$

where $\hat{h}$ is a unit vector tangential to the surface. This boundary is called "ideal" since the complex Poynting vector $\bar{S} = \frac{1}{2} \bar{E} \times \bar{H}^*$ has no component normal to the boundary on the surface. This can be seen through the expansion

$$\begin{aligned} \hat{n} \cdot (\bar{E} \times \bar{H}^*) &= [\hat{h} \times (\hat{n} \times \hat{h})] \cdot (\bar{E} \times \bar{H}^*) \\ &= (\hat{h} \cdot \bar{E})[(\hat{n} \times \hat{h}) \cdot \bar{H}^*] - [(\hat{n} \times \hat{h}) \cdot \bar{E}](\hat{h} \cdot H)^* \\ &= 0 \end{aligned} \tag{18}$$

where $\hat{n}$ is the unit vector normal to the SHS. In a similar way, it can be proven that the complex Poynting vector has a zero component in the direction (say $\hat{s}$) transverse to $\hat{v}$ on the SHS. This means that the SHS in direction $\hat{s}$ can be considered as a soft surface, a concept that originated from acoustics. One can take advantage of the characteristics of the soft surface to block the surface waves propagating outward along the high dielectric-

73

constant substrate, thus alleviating the diffraction at the edge of the substrate. Hence, a patch antenna with arbitrary configuration lies on a finite substrate with a dielectric constant of ε_r and is surrounded with an SHS that is formed as a soft surface in the outward direction. As a result, it would be difficult for surface waves to propagate from the microstrip patch to the substrate edge. Such an SHS can be realized using a number of via rings whose height, h, must be approximately equal to $\lambda_0/(4*\varepsilon_r^{1/2})$, where λ_0 is the free space wavelength. Based on this basic configuration, the author investigates the design of a patch antenna on a large-size substrate surrounded by SHS rings in the next two sections of this chapter.

5.2 Antenna Structure

The antenna structure for this implementation is shown in Figure 43. It consists of two square patches that are stacked for bandwidth enhancement. Each side of the patches is 0.75 mm. The LTCC substrate has a dielectric constant, ε_r, of 5.4 and a loss tangent of 0.0015. The substrate thickness is 500 μm composed of five total layers that are each 100 μm thick. The upper patch is placed on the top layer of the substrate while the lower patch is positioned three layers below the upper patch. There is one layer that separates the intermediate ground plane and the lower patch. A 100 μm cavity has been etched out between the intermediate ground plane and the total ground plane of the structure for the purpose of burying the MMICs. Surrounding the patch antennas are square metallic rings that mimic the perfect electric conducting (PEC) regions of the SHS. These rings are separated by a distance of 0.5 mm. This distance is chosen arbitrarily. Although a smaller distance does not have a significant effect on the suppressing ability of the SHS, a larger distance allows the fields to continue to propagate to the edges of the

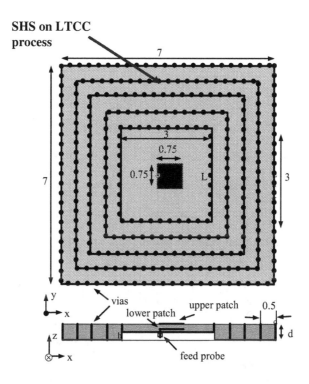

Figure 43. Stacked patch antenna surrounded by SHS structure.

substrate. The substrate that is between the rings represents the perfect magnetic conducting (PMC) regions of the SHS. In simulation, continuous metal walls can be used for the PEC regions, while in fabrication, it is necessary to use vias that are separated by 0.5 mm. The diameter of the vias is 0.13 mm. A requirement for an effective, surface wave suppressing SHS is that the substrate thickness maintains a height of $\lambda_o/ (4*\varepsilon_r^{1/2})$, where λ_o is the free space wavelength. Because of this limitation, the resonant frequency for this design is chosen to be 64.55 GHz. The metal for the conductors is silver (Ag). The total area for the antenna design is 7 mm x 7 mm.

5.3 Simulation Results and Near Field Analysis of SHS Antenna Structure

The antenna of Figure 43 was simulated using MicroStripes 5.6. A similar design without the SHS structure was also simulated in the software to compare the performance of the two structures. The total size of both designs is 7 mm x 7mm x 0.5 mm. The radiation patterns at the resonance are shown in Figure 44. Two major effects can be seen from the E-plane and H-plane radiation patterns. The first is the reduction in the backside radiation (lower hemisphere of the plots) which is less than 5 dB below the main beam for both patterns in the design without the SHS. In contrast, the design with

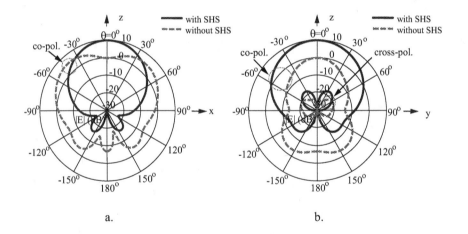

a.   b.

Figure 44. 2D radiation patterns on a.) E-plane and b.) H-plane for designs with and without SHS

the SHS exhibits F/B ratios around 15 dB in the E- and H-planes. The cross-polarization level, which can be seen in the H-plane, is slightly higher in the SHS implemented structure compared to the structure without the SHS, but its value (below -18 dB) is suitable for this design. The second effect that can be seen in the two plots is the increased gain in the design with the SHS. Both the E-plane and H-plane radiation patterns exhibit on-axis ($\theta=0°$) gain enhancements of 10 dB. The major reason for this enhancement is due to the edge diffraction of the finite-size substrate. As the finite-size substrate increases, the edge diffraction effects contribute increased radiation at an angle between 30-45° away from the broadside direction (normal to the surface of the patch). This is only seen in the E-plane plot in which the resonant length is along the horizontal axis.

The return loss versus frequency plots for the two designs are shown in Figure 45. It is observed that one TM_{10} resonance is present in the structure at the resonant frequency of 64.55 GHz. The impedance bandwidth of the SHS implemented structure is smaller than that of the design without the SHS. This is due to the contribution of the currents on the metals that adds inductance to the antenna. The on-axis gain versus frequency is also displayed in Figure 45. It can be observed that the gain of the antenna with the SHS is around 10 dB and this value is constant over the entire bandwidth from approximately 61.5-66 GHz. The on-axis gain for the antenna structure without the SHS is 0 dB.

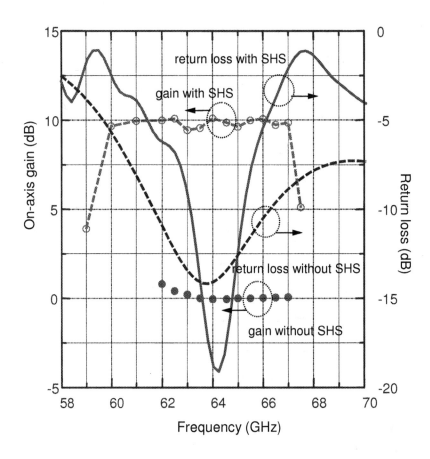

Figure 45. Return loss and on-axis gain versus frequency for antenna designs with and without SHS.

Plots of the near-field distributions for both designs are presented in Figure 46. As the fields propagate away from the radiating edges of the patch, significant field intensities are present at the edges of the substrate in the design without the SHS. On the other hand, when the SHS is implemented to the patch antenna design, the electromagnetic fields become more and more weak at they propagate toward the SHS. At the edges of the substrate, the fields are significantly suppressed. Due to the results of the near-field distributions, it is believed that an SHS structure using a smaller number of rings could be used to obtain the same effect of suppressing substrate modes as they propagate away from the radiating edges of the patch. This will also reduce the overall size of the structure, thus allowing for a more compact device for integration to RF SOP modules.

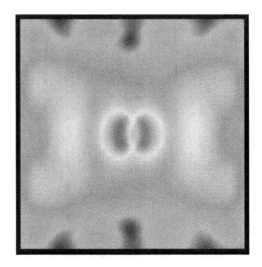

a.

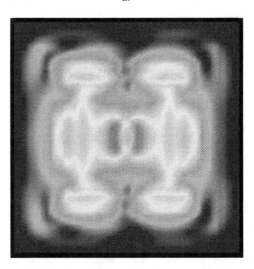

b.

Figure 46. Near-field distributions on electric field for antenna design a.) without SHS and b.) with SHS.

The concept of using an SHS structure to effectively suppress the surface waves that propagate towards the edges of high dielectric constant ($\varepsilon_r > 5$) substrates has been discussed. In this section of the chapter, a second implementation of an SHS structure that utilizes the soft surface properties is applied to improve the radiation pattern of a patch antenna. The major difference in this implementation is the use of only one soft surface ring as opposed to the five rings that were previously needed for effective surface wave suppression and edge diffraction elimination. In addition, the substrate thickness requirement of a guided quarter-wavelength, $\lambda_0/(4*\varepsilon_r^{1/2})$, can be removed; therefore, various thicknesses of substrates can be used at a given frequency. This design can also be easily realized in fabrication by using vias in the substrate layers and square metalized rings on the metal layers.

5.4 Investigation of Improved One Ring Soft Surface Structure and its Implementation to a Patch Antenna

An illustration of an antenna structure surrounded by a single soft surface ring is shown in Figure 47. Since the focus of this section is on the operation of the ring, a simple square patch antenna is used as the radiator for simplicity. The ideal soft surface ring is shorted to ground along its outer edge with a continuous metal wall. The important parameters of the ring are its width, W_s, and length, L_s. It is seen in Figure 47 that the ring can be modeled as a transmission-line where the impedance at the shorted edge is zero. Additionally, it is assumed that the inner edge of the ring can be modeled as an open circuit whose impedance is infinite.

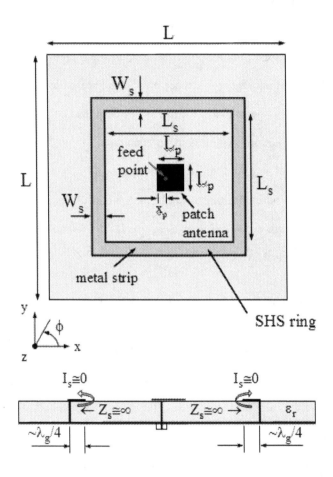

Figure 47. Patch antenna surrounded by a single soft surface ring.

Therefore, to effectively transform a short circuit to an open circuit, the width, W_s, must be approximately equal to a guided quarter-wavelength,

$$W_s^* = \frac{c}{4 f_0 \sqrt{\varepsilon_r}},$$ (19)

where c is the speed of light ($\approx 3*10^8$ m/s), f_0 is the operating frequency, and ε_r is the dielectric constant of the medium. Taking into account the fringing fields, W_s can be more accurately expressed as

$$W_s = \frac{c}{4 f_0 \sqrt{\varepsilon_{eff}}} - \Delta W_s.$$ (20)

Equations for the fringing field width, ΔW_s, and the effective dielectric constant, ε_{eff}, can be obtained in [11] as

$$\Delta W_s = H \cdot \left[0.882 + \frac{0.164(\varepsilon_r - 1)}{\varepsilon_r^2} + \frac{\varepsilon_r + 1}{\pi \varepsilon_r} \times \left\{ 0.758 + \ln\left(\frac{2W_s^*}{H} + 1.88 \right) \right\} \right]$$ (21)

and

$$\varepsilon_{eff} = \left[\frac{\varepsilon_r + 1}{2} + \frac{\varepsilon_r - 1}{2} F(W_s^* / H) \right]$$ (22)

where $F(W_s^*/H) = (1 + 6H/W_s^*)^{-\frac{1}{2}}$ and H is the thickness of the substrate. The design value for W_s is optimized by simulation for a maximum directivity in the broadside ($\theta = 0°$) direction.

In order to demonstrate the effectiveness of the compact soft surface ring for radiation pattern improvement and analyze the length, L_s, simulations were performed of the structure in Figure 47 for three dielectric constant values ranging from low to high values (ε_r = 2.9, 5.4, and 9.6). These values correspond to typical dielectric constants of materials such as LTCC and LCP. The operating frequency is 15 GHz, while the size and thickness of the substrate are fixed at 40 mm x 40 mm ($2\lambda_0$ x $2\lambda_0$) and 0.5 mm ($0.025\lambda_0$), respectively. The length of the patch, L_p, and the width of the ring, W_s, is varied accordingly to maintain resonance at 15 GHz for the different dielectric constant values. The values of the lengths are 5.40, 3.94, and 2.90 mm and the ring widths are 2.65, 1.88, and 1.36 mm for ε_r = 2.9, 5.4, and 9.6, respectively. Figure 48 illustrates how the directivity at broadside is varied as a function of the soft surface length, L_s.

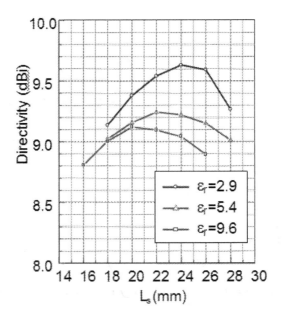

Figure 48. Directivity versus soft surface length, L_s, for three dielectric constants.

It is observed that directivities greater than 9 dBi can be obtained for all three dielectric constant values when the compact soft surface ring is employed. Simulations were also performed for a patch antenna with the same substrate size and thickness dimensions stated previously, but without the soft surface ring. Because of the edge diffraction effects that arise when a large finite substrate is used, a dip in the broadside direction of the co-polarized component of the E-plane radiation pattern is present, and hence, the broadside directivity is suppressed. Conventional patch antennas can achieve directivities up to 7 dBi. The directivities for the patch antennas without the ring are 5.7, 4.5, and 4.0 for $\varepsilon_r = 2.9$, 5.4, and 9.6, respectively. Based on these results, it is observed that the soft surface ring can also increase the directivity above the expected value by about 2 dBi. There is a strong correlation between the soft surface length and the directivity. It can be clearly seen that an optimal value exists for this length in order to achieve maximum broadside directivity. The maximum directivities at $L_s = 24$, 22, and 20 mm are 9.6, 9.2, and 9.1 dBi for $\varepsilon_r = 2.9$, 5.4, and 9.6, respectively. An explanation of the optimal soft surface length is discussed later.

A plot of the directivity as a function of frequency around 15 GHz for the design with the ring is displayed in Figure 49. Directivities above 9 dBi can be observed around this small frequency band. The directivities for $\varepsilon_r = 2.9$ exhibit less variation (directivities > 8 dBi between 14.5-15.4 GHz) than those of the designs with $\varepsilon_r = 5.4$ and 9.6. In fact, there is a large decrease (as much as 2.5 dBi) above and below 15 GHz; hence, the bandwidth of directivity greater than 9 dBi is 4 % for $\varepsilon_r = 2.9$, but for $\varepsilon_r = 5.4$ and 9.6, the bandwidths are around 2%.

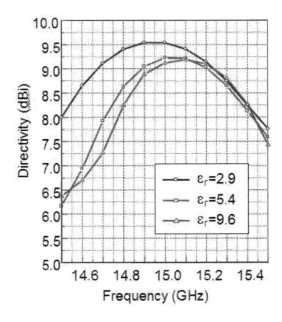

Figure 49. Directivity versus frequency for three dielectric constants.

The return loss versus frequency is shown in Figure 50 for a patch antenna with and without the ring for $\varepsilon_r = 2.9$. It can be observed in the simulated plot that the soft surface ring does not significantly affect the return loss. This is primarily due to the placement of the ring which is approximately a half-wavelength away from the edges of the patch. The same observation exists for the return loss plots when $\varepsilon_r = 5.4$ and 9.6.

The radiation pattern comparison between the patch structure with and without the ring at 15 GHz for $\varepsilon_r = 9.6$ is displayed in Figure 51. In the E-plane radiation pattern, the co-polarized component of the design without the ring experiences a smaller broadside directivity and increased areas of radiation around $\pm 30°$ due to the surface wave propagation toward the edges of the substrate and the edge diffraction effects of using a

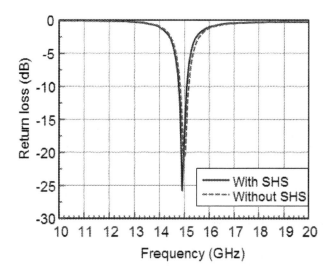

Figure 50. Return loss versus frequency for $\varepsilon_r = 2.9$.

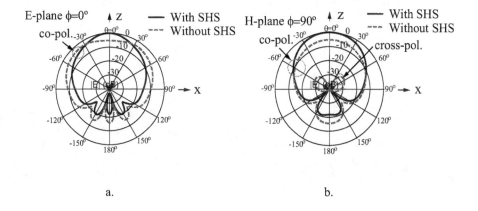

a.   b.

Figure 51. a.) E-plane and b.) H-plane radiation patterns of patch antenna surrounded by soft surface ring.

large finite-size substrate. When the ring is introduced, the maximum radiation is enhanced in the broadside direction. In addition, it is observed in the H-plane patterns that the gain is significantly increased (above 5 dBi), and the cross-polarization is reduced. In addition, the backside radiation level is decreased by about 8 dB. The electric field distributions on the top surface of the substrate for the antenna design with and without the ring are shown in Figure 52. As the electric field decreases away from the radiating edges of the patch, the ring contains that field so that the effective aperture area is small and energy does not continue to propagate toward the edges of the substrate. Without the ring, there is no mechanism in place that stops the energy from propagating away from the antenna.

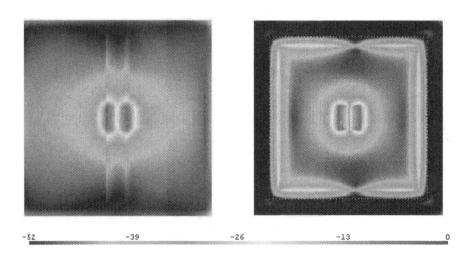

| -52 | -39 | -26 | -13 | 0 |

a.   b.

Figure 52. Electric field distribution in the near field for design a.) without and b.) with the soft surface ring.

There are two factors to consider when evaluating the cause of radiation pattern improvement from this compact structure. First, the short-circuited, quarter-wave metalized strip serves as an open circuit to the TM_{10} mode (the dominant mode of the patch). This is because the width of the strip is designed to operate at the same frequency as the antenna. Therefore, it is difficult for the surface current on the inner edge of the soft surface ring to flow outward. As a result, the surface waves are significantly suppressed outside the ring. The second factor is the modeling of the fringing fields of the ring and those of the patch as a two-element array in the E-plane on each side of the antenna. Figure 53 is an illustration of the fringing field effect that forms the array. This model is valid when the distance between the inner edge of the ring and the edge of the patch is approximately a half-wavelength in free space ($\lambda_0/2$). Even though, the magnitude of the fringing field of the soft surface may be much lower compared to that of the radiating edges, the size of the ring is much larger than the patch. Hence, the contribution of the broadside radiation of the ring is large enough to form an array with the antenna. When L_s is increased, the fringing field will become much weaker and the operation of the ring will be insignificant. As a result, the optimal value for L_s is given as

$$L_s \approx \lambda_0 + L_p . \tag{23}$$

In addition, it has been shown via simulation that adding more rings does not improve the radiation pattern performance and contributes to an increase in the size of the structure.

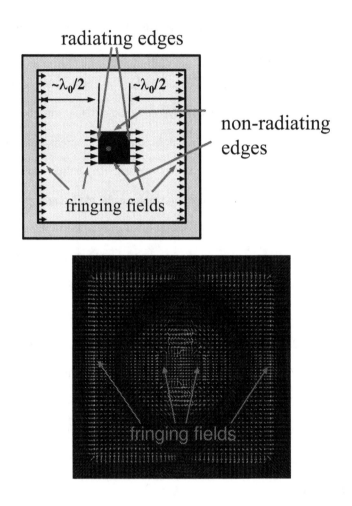

Figure 53. Fringing field effect of patch antenna and soft surface ring.

5.5 Measurement Results of Practical One Ring Soft Surface Implementation to a Patch Antenna

The design concepts that were formulated in the last section are applied using LTCC multilayer technology. Based on limitations in the fabrication process, the thickness of the substrate is fixed at 1100 μm. This consists of 11 total substrate layers that are each 100 μm thick. The dielectric constant is 5.4 and the loss tangent is 0.0015. The design frequency for this fabricated module is 16.5 GHz (applicable in the Ku band). The length, L_p, is 3.4 mm, and the length and width (L_s and W_s) of the soft surface ring are 22.2 mm and 1.7 mm, respectively. Figure 54 shows an illustration of the antenna structure. Based on the fixed substrate thickness, a stacked configuration is employed to improve the bandwidth and eliminate the feed inductance. In simulation, a coaxial probe is used to excite the antenna, but in fabrication, the design is excited by a microstrip line that is connected to the patch by a via through a small hole in the ground plane. The microstrip line is placed on the bottom side of the module. A ground plane rests 200 μm (two layers) above the microstrip line. The microstrip line is placed below the ground plane to alleviate its contribution of spurious radiation to the radiation pattern and to avoid its interference with the soft surface ring. Another 200 μm (two layers) above the ground lies the lower patch (excited directly with a via). Finally, the upper patch (excited through capacitive coupling) is placed on the top side of the module. The distance between the lower and upper patch is 700 μm (seven layers). The total lateral size of the LTCC board is 30 mm x 30 mm. The diameter of the vias that short-circuit the rings to ground is 100 μm, and the center-to-center spacing between the vias is 300 μm.

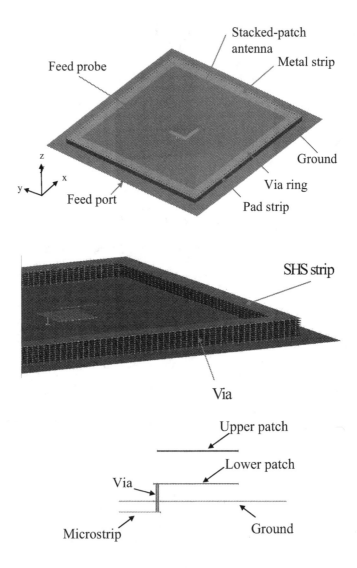

Figure 54. Practical illustration of improved soft surface ring surrounding a patch antenna.

Figure 55 displays the fabricated antenna with small pads for coplanar waveguide (CPW) probe measurement of the return loss and larger pads for radiation pattern measurement using a SMA coaxial connector. To compare the results of the antenna's radiation pattern with and without the ring, a simple stacked patch antenna with the same dimension, layer placement and method of excitation was also fabricated and measured along with the ring.

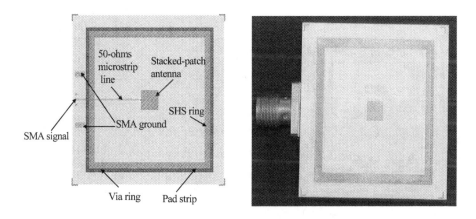

Figure 55. Layout and fabricated picture of improved soft surface ring surrounding a patch antenna.

The simulated and measured return loss versus frequency for the designs with (and without) the soft surface ring are presented in Figure 56. A good agreement is observed between the simulated and measured plots of both designs. The return loss is more sensitive to the inclusion of the soft surface ring in the stacked configuration because the ring acts as a radiator as well, but the impedance performance is dominated by the excitation of the stacked elements. Nevertheless, the bandwidth of the both designs is improved with the incorporation of the second patch. The measured bandwidth at -8 dB is around 9% for the design with the ring. The measured bandwidth at -10 dB is split into two bands of the two resonance modes (generated by the lower and upper patches). This discrepancy could be due to some fabrication issues such as the variation of the dielectric constant or some effects that exist from the transition between the coaxial connector and the microstrip line. In addition, there is a 300 MHz shift in the lower frequency resonance which could be due to the variation of the material's dielectric constant. It is important to note that the dielectric constant of many practical substrates tends to have a tolerance of ±2%.

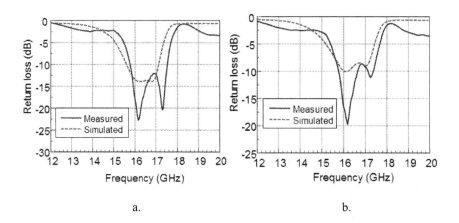

a.   b.

Figure 56. Simulated and measured return loss versus frequency for design a.) without and b.) with the soft surface ring.

The radiation patterns of the simulated and measured designs are displayed in Figure 57 for the designs with (and without) the ring. The measurements were performed at 17 GHz. A good agreement is observed between the simulated and measured co-polarized patterns in the E- and H-plane. The backside radiation of the design with the ring is about 5-6 dB lower than that without the ring. The measured cross polarization is slight higher than that observed in simulation due to a slight misalignment of the antenna away from the E- and H-plane. In simulation, it is observed that higher cross-polarization can occur at $\varphi=45°$ and 135°, so slight misalignments can increase this polarization. Finally, the broadside ($\theta=0°$) gain of the antenna with the ring (9 dBi) is greater than 6 dBi larger than the design without the ring (2.5 dBi). The simulated efficiencies of the all designs considered in this analysis are above 88%. In comparison to a patch antenna surrounded by a PBG structure of similar size, the F/B ratio of the soft surface ring is close to 10 dB higher than a design that incorporates the PBG.

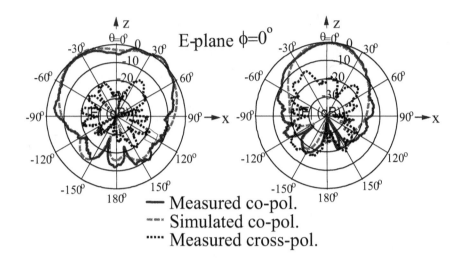

a.

b.

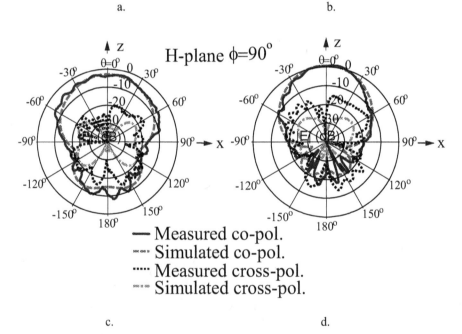

c.

d.

Figure 57. Radiation patterns of E-plane components a.) without and b.) with the ring along with H-plane components c.) without and d.) with the ring of the practical design.

5.6 Sensitivity Analysis and Optimization Using Design of Experiments and Monte Carlo Simulation

The performance of RF and microwave systems is severely affected by tolerances in the fabrication process, especially for small circuit dimensions at microwave frequencies. Therefore, it is important to analyze the performance of the system (in this case, the antenna) by examining the most critical factors that affect the performance of the soft surface ring. The performance is modeled with transfer functions developed from a methodology based on the integrated use of statistical tools, deterministic simulations, and measurements. The statistical tools used in the methodology include sources of variation tools such as Analysis of Variance (ANOVA), Statistical Process Control (SPC), and Monte Carlo simulation combined with Design of Experiments (DOE) tools such as 2^k designs and Response Surface Methodology (RSM) for the development of the transfer functions.

Because the fabrication of the test structures is expensive and time consuming, which is very common especially for 3D architectures, the sensitivity analysis is based on a combination of electromagnetic and statistical simulations. The first step of the methodology is to statistically quantify the nominal values and the variation for the two design parameters that are achievable by current fabrication processes. The next steps include the development of the transfer functions that utilize statistically quantified design parameters to predict the nominal values and process-based variations of the three figures of merit for system performance. These figures of merit are the directivity (D), the F/B ratio, and the maximum return loss peak (MRLP) between the two resonances. It is optimal to keep the MRLP below -10 dB in order to maintain an impedance bandwidth that encompasses both resonances (not just a single resonance). In addition, it is common to examine the bandwidth at -10 dB where 90% of the input power enters the excitation port of the antenna.

The transfer functions are developed using Design of Experiments. The experimentation method chosen for the DOE is a full factorial design with center points [43]. The factorial designs are used in experiments involving several factors where the goal is the study of the joint effects of the factors on a response. First, the variables that affect this design are chosen to be the length, L_s, and width, W_s, of the soft surface ring. The analysis intervals for the two variables are presented in Table 1.

Table 1. Ranges for the soft surface ring input variables.

	L_s (mm)	W_s (mm)
"-" level	12	0.5
"+" level	36	4.5
Center point	24	2.5

Center points in the design increase the capability of investigating the model fit, including curvature in the response, and account for variation in the fabrication process of the structure. Since the statistical models are based on deterministic simulations, the variation of the center points were statistically simulated based on a ± 10 µm tolerance and a 3σ fabrication process for all inputs, given by the X and Y shrinkage factor for the LTCC substrate.

The statistical analysis of the first order models shows which effects and interactions between the factors are significant for each of the three figures of merit, and those that are not significant are eliminated from the final models. In this case, curvature has not been detected and the first order models, which are investigated for the normality and equal variance assumptions, are presented below:

$$D=6.72-0.29\left(\frac{L_s-24}{12}\right)+0.09\left(\frac{W_s-2.5}{2}\right)-\left(\frac{L_s-24}{12}\right)\left[0.33\left(\frac{W_s-2.5}{2}\right)\right], \quad (24)$$

$$F/B=9.96-0.80\left(\frac{L_s-24}{12}\right)+0.55\left(\frac{W_s-2.5}{2}\right)-\left(\frac{L_s-24}{12}\right)\left[0.85\left(\frac{W_s-2.5}{2}\right)\right], \quad (25)$$

and

$$MRLP= -9.76-0.59\left(\frac{L_s-24}{12}\right). \quad (26)$$

Next, the structure is optimized based on the models for the following goals: maximum directivity, maximum F/B ratio and a minimum MRLP. The optimal values are found to be 7.07 dBi, 11.2 dB, and -9.5 dB, respectively, for the combination of inputs, L_s = 19 mm and W_s = 4.5 mm. This is an interesting result considering that the directivity and the F/B ratio are considerably larger in the practical design (9.2 dBi and 15 dB, respectively). It is evident that the goal of minimizing the MRLP has a significant affect on degrading the performance of the directivity and F/B ratio. This can be attributed to the fact the three figures of merit were weighted evenly. Therefore, based on the need to minimize the MRLP, the directivity and F/B ratio may suffer a major decrease in the value obtained via electromagnetic simulation and measurement. Use of these optimization tools suggest that a -8 dB impedance bandwidth is optimal in maintaining a high directivity and a high F/B ratio, but this is predicated on the specifications of the application. Additionally, the optimized width, Ws, does not coincide with the theoretical width, $\lambda_0/(4*\varepsilon_r^{\frac{1}{2}})$, that is necessary to obtain an optimal performance. This is another reason why the directivity and F/B ratio is smaller than expected based on electromagnetic simulations and measurements. In the future, additional optimization simulations will be performed to examine the performance of the directivity and F/B ratio when the MRLP is minimized to -8.5 or -9 dB.

Lastly, the performance capability of the system is evaluated for the optimal structure using Monte Carlo simulation [44]. The Monte Carlo simulation of 1000 trials provides evidence that the given specification limits yield long term six sigma process capability. Table 2 shows the results of the Monte Carlo simulation.

Table 2. Predicted performance of the outputs of the soft surface ring design.

	D (dBi)	F/B (dB)	MRLP (dB)
Nominal	7.07	11.2	-9.5
USL	n/a	n/a	-8.756
LSL	6.709	9.919	n/a
Cp	n/a	n/a	n/a
Cpk	1.57	1.52	1.58

The first row shows the nominal values of the outputs obtained by plugging in the optimized values of the inputs into the models. USL and LSL are the upper and lower specification limits, respectively, and they represent the worst case scenario for each of the outputs. It is noted that the outputs that have been optimized for a maximum directivity and F/B ratio and a minimum MRLP, so the directivity and F/B ratio have a value for LSL, but no value for USL. Conversely, the MRLP has a value for USL, but no value for LSL. Cp and Cpk are metrics that quantify evidence that the system complies with six sigma process capability. Six sigma capability is reached for processes that achieve Cp ≥ 2 and Cpk ≥ 1.5 for processes with USL and LSL, and Cpk ≥ 1.5 for processes with only USL or LSL, allowing, in both cases, the possibility of a long term ±1.5 sigma shift. In this case, Table 2 shows that these conditions are satisfied and evidence that six sigma capability, which includes the possibility of a long term ±1.5 sigma shift, is reached. In other words, the designer has knowledge at the beginning of the design process that approximately 3.4 measurements out of 1,000,000 may occur beyond these specification limits (i.e., the USL and LSL). If the designer finds these limits to be unacceptable, the whole system can be redesigned to achieve the desired specification limits without the need of building any test structures; hence, an expensive and time consuming design cycle can be alleviated.

CHAPTER 6

APPLYING LUMPED ELEMENT EQUIVALENT CIRCUITS TO THE DESIGN OF SINGLE-PORT MICROSTRIP ANTENNAS AND RESONANT STRUCTURES

The continuing growth in the design and use of microstrip antennas and resonator circuits for a wide range of shapes, integrated modules, and applications has necessitated the need for increased research to fully interpret the physics behind the structures. The use of computational methods, such as the finite difference time domain (FDTD) [45], the method of moments (MoM) [46], the finite element method (FEM) [47], and the transmission-line matrix (TLM) method [48], has aided in the understanding of microstrip antennas from the electromagnetics viewpoint. These computational methods have been integrated into commonly used computer-aided design (CAD) packages, such as High Frequency System Simulator (HFSS) [49], Sonnet Suite [50], and MicroStripes [51]. To take full advantage of analyzing resonant structures and patch antennas, it is necessary to couple these computational methods with circuit analysis techniques that can potentially provide researchers with additional information to explain phenomena (such as higher-order modes and parasitics) that may be present due to lossy substrates, complex discontinuities, or metal surfaces. Design equations for analyzing microstrip antennas based on the parallel resistor-inductor-capacitor (RLC) circuit representation have been presented in [11]. Additionally, Pozar has explained how microstrip circuits can be represented by open-circuited half-wavelength ($\lambda/2$) transmission-lines [52]. The scattering (S-) parameter data from these circuit representations do not always agree with the data obtained from the computational full-wave electromagnetic methods. Hence, there is a need for improved circuit representations to bridge the gap between computational and circuit analysis, while accelerating the design process.

One particular equivalent circuit that has been used for the modeling of microwave topologies is derived from the use of rational function approximations from a vector fitting approach. It consists of utilizing passive lumped elements; the values and the number of passive lumped elements are based on the nature of the poles and residues, the order of approximation, and the number of iterations. In the case of N-port networks, where N≠1, transformers are incorporated into the analysis. In [53], the authors used this approach to model a Wilkinson power divider at 4 GHz. Furthermore, Araneo and Celozzi have successfully modeled microwave discontinuities, such as the open-ended microstrip line, the via, and the T-junction power divider by vector fitting [54-55]. Finally, the authors in [56] proposed an equivalent circuit for the analysis and modeling of linear dipole antennas for UWB applications, while the authors in [57] attempted to apply this procedure to a patch antenna without success because the circuit representation produced negative values in the resistance. Based on research applying this method to antennas, this technique seems to succeed well for fundamental microstrip antenna structures in which a low order ($\leq 4^{th}$) approximation can be employed. More complex structures with many resonances (fundamental and parasitic) require a much higher order ($> 8^{th}$) approximation to accurately model the scattering parameters, and hence, a circuit with many parameters has a higher probability of non-physical negative resistive values. For the first time, a simple method for addressing the issue of unphysical negative resistances is applied to the design and analysis of single-port TM_{10} mode microstrip antennas and other resonant structures for the extraction of equivalent circuits, and the circuit models agree very well with full-wave electromagnetic simulation software.

6.1 Approximating Rational Functions and Passivity Enforcement

The values of the lumped passive components for an equivalent circuit are determined through the analysis of approximating rational functions by vector fitting. In this section, the author felt it is necessary to include a brief discussion on how this method works. More complete information on "vector fitting"-based approximations is given in [58].

The first step in this process is to acquire the scattering, impedance, or admittance parameter data, acquired via full-wave electromagnetic simulation (for one design) versus frequency. There is no rule for the number of frequency points that should obtained, but it is suggested to use no fewer that 100. Based on the lumped network circuit model used in this paper that is derived from a single-port admittance and to maintain consistency with previously published articles, the scattering or impedance parameter data has to be transformed to admittance parameter data. The author wanted to stress the importance of using this method for single-port antennas because many antenna applications require single feed points. Then, a rational function approximation, $y_{fit}(s)$, is used to approximate the admittance parameter data. This function is shown below:

$$y_{fit}(s)=\sum_{n=1}^{N} \frac{c_n}{s-a_n}+d+se \qquad (27).$$

where s is a single frequency point ($= j\omega$), c_n and a_n are the residue and pole values, respectively from n=1,2,...,N where N is the number of poles (order of approximation), and d and e are higher order coefficients. This equation can be further decomposed into an equation that separates the real poles from the complex poles. The equation:

$$y_{fit}(s)=\sum_{n=1}^{N_r}\frac{c_n}{s-a_n}+\sum_{n=1}^{N_c}(\frac{\tilde{c}_n}{s-\tilde{a}_n}+\frac{\tilde{c}_n^*}{s-\tilde{a}_n^*})+d+se \qquad (28)$$

consists of two summations, one for the real poles and residues and the other for the complex poles and residues. In equation 28, N_r represents the number of real poles, while c_n and a_n are the real residue and pole values, respectively. In the second summation, N_c represents the number of complex pole pairs, while $\tilde{c}_n$ represents a complex residue (with its complex conjugate, $\tilde{c}_n^*$) and $\tilde{a}_n$ represents a complex pole (with its complex conjugate, $\tilde{a}_n^*$). In the beginning, equation 27 is nonlinear since the unknown poles, a_n, are in the denominator. To convert this function into a linear one that could be solved with straightforward techniques, the poles are transformed into known quantities by giving them values at the beginning of the fitting process. These are called starting poles, denoted by $\bar{a}_n$, and they only exist at the start of the evaluation. Equation 29 is a representation of equation 27 called $(\sigma y)_{fit}(s)$ where the poles are represented by $\bar{a}_n$. This is shown below:

$$(\sigma y)_{fit}(s)=\sum_{n=1}^{N}\frac{c_n}{s-\bar{a}_n}+d+se \qquad (29)$$

where again, $\bar{a}_n$ are the starting poles in the data fitting process. Notice that all the parameters are the same except the poles. To effectively use the vector fitting process and create a linear set of equations with the same unknowns (as opposed to nonlinear), one must select a set of starting poles. In smooth functions, the poles must exist in the frequency range of interest, but there are no limitations as to where the poles must lie. In non-smooth plots, the poles are selected based on the frequency at which minimum and maximum peaks exist in the admittance data [58]. The number of poles that are used is determined by the number of peaks that are present in the data. Gustavsen and Semlyen suggest that at least one complex pole pair should be used per peak. When the admittance data is smooth without peaks, the starting poles can be real. For example, if peaks exist at 4.2, 4.5, and 4.7 GHz in the admittance plot of a simulation between 4 to 5 GHz, then a good set of poles to choose for starting poles would be $-4.2e7 \pm j*4.2e9$, $-4.5e7 \pm j*4.5e9$ and $-4.7e7 \pm j*4.7e9$. Three points are evident in the selection of these poles. First, the complex part of the pole corresponds to the frequency at which the peak exists. Second, the real part of the pole is negative. Lastly, the real part is 100 times smaller than the imaginary part. The first point has been explained. The second point is enforced to maintain stability (poles must exist on the left-half plane of a complex plot). The last point will be discussed later. A rational approximation for equation 28, denoted $\bar{\sigma}(s)$, is shown below:

$$\bar{\sigma}(s) = \sum_{n=1}^{N} \frac{\bar{c}_n}{s - \bar{a}_n} + 1 \tag{30}$$

where $\bar{c}_n$ are the residues different from c_n. Multiplying equation 30 with $y(s)$ ($y(s)$ represents the corresponding admittance value at a single frequency point, s, acquired during the simulation) and equating it with equation 28, gives $(\sigma y)_{fit}(s) = \bar{\sigma}(s) \cdot y(s)$. Substituting in the equations for $(\sigma y)_{fit}(s)$ and $\bar{\sigma}(s)$ gives equation 31 shown below:

$$\sum_{n=1}^{N} \frac{c_n}{s-\bar{a}_n} + d + se = (\sum_{n=1}^{N} \frac{\bar{c}_n}{s-\bar{a}_n} + 1) \cdot y(s) \qquad (31)$$

Then, y(s) is multiplied to the terms in $\bar{\sigma}(s)$ and isolated by itself to obtain the following

equation:

$$(\sum_{n=1}^{N} \frac{c_n}{s-\bar{a}_n} + d + se) - (\sum_{n=1}^{N} \frac{\bar{c}_n \cdot y(s)}{s-\bar{a}_n}) = y(s) \qquad (32)$$

Solving y(s) for every single frequency point leads to a matrix equation, Ax=b, where A

is row vector of dimension 2(N+1) relating known terms s, a_n, $\bar{a}_n$, and y(s), x is a

2(N+1)-dimensional column vector of unknown terms, c_n, $\bar{c}_n$, d, and e, and b is a scalar

of the known y(s) value. To set up the fitting for a vector of frequency points, Equations

31 and 32 are repeated for each frequency point leading to additional rows of A; hence, A

is a P x 2(N+1)-dimensional matrix where P is number of frequency points in the

simulation, x remains a 2(N+1)-dimensional column vector of the unknown terms, and b

becomes a P-dimensional column vector of y(s) values (each one corresponding to a

frequency point, s, in simulation). This leads to the necessity of having the real part of

the complex pole be significantly smaller that the imaginary part in order to prevent an

ill-conditioning of the linear problem, Ax=b. The layout of the matrices is shown in

Appendix A of [58]. The matrix equation Ax=b can be solved by using the singular

value decomposition (SVD) method. After the elements in the column vector x are

solved in the matrix equation, $(\sigma y)_{fit}(s)$ and $\bar{\sigma}(s)$ can be written as:

$$(\sigma y)_{\text{fit}}(s)=e \cdot \frac{\displaystyle\prod_{n=1}^{N+1}(s-z_n)}{\displaystyle\prod_{n=1}^{N}(s-\bar{a}_n)} \qquad (33) \qquad \text{and} \qquad \bar{\sigma}(s)=\frac{\displaystyle\prod_{n=1}^{N}(s-\bar{z}_n)}{\displaystyle\prod_{n=1}^{N}(s-\bar{a}_n)} \qquad (34)$$

where solving for $y(s)$ gives $y(s) = (\sigma y)_{\text{fit}}(s)/\bar{\sigma}(s)$. Now it can be seen that the new poles, a_n, for $y(s)$ are the zeros, $\bar{z}_n$, of $\bar{\sigma}(s)$. Now $y(s)$, shown in the form of a ratio of products, can be transformed into a sum of ratios to look like the form in equation 27. If an additional iteration of approximating the function is desired, the zeros, $\bar{z}_n$, of $\bar{\sigma}(s)$, become the new starting poles, $\bar{a}_n$, of $(\sigma y)_{\text{fit}}(s)$ and $\bar{\sigma}(s)$ and the fitting is repeated. This process can be continued in an iterative fashion until the parameters in $y_{\text{fit}}(s)$ are properly chosen to fit the admittance data because the selection of starting poles are no longer necessary since the zeros of $\bar{\sigma}(s)$ are used.

There is an important critical step that must be taken into account to ensure that passivity is maintained in the fitting process. The necessity of maintaining passivity is associated with the desire to eliminate (spurious) reactive power. In N-port networks, the eigenvalues of the conductance matrix, G_{fit}, at all frequencies must be forced to positive values [59]. This criterion is simpler in single-port networks where only the conductance, G_{fit}, at all frequencies needs to maintain a positive value.

After the final c_n, a_n, d, and e parameters have been obtained, these values are correlated to an equivalent circuit shown in Figure 58. Design equations for transforming these parameters to the lumped elements of Figure 58 are presented in [59] and [60], while the concept of forming circuits from rational factors is shown in [61]. The R_r-L_r

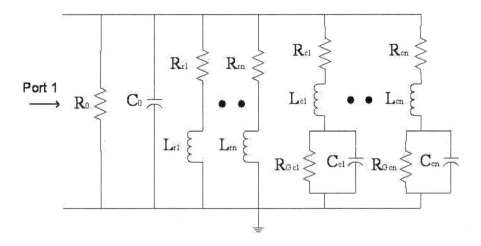

Figure 58. Schematic of single-port equivalent circuit representation.

branch corresponds to a real pole, while the R_c-L_c-R_{Gc}-C_c branch represents a complex pole pair.

6.2 Resistance Invertibility

In order to transform equivalent circuits with negative resistance values to physical circuits, a new method is proposed. This method involves inverting the negative resistances to positive values. The circuit element that is used to determine whether this method can be done is the value of R_c. If the magnitude of this value is less than around 100 Ω, then both R_c and R_{Gc} can be inverted to produce an accurate physical circuit. This is a safe approximation based on analysis of generating circuits with other values of R_c, in which the circuit did not correlate well with the approximation. After analyzing the equivalent circuits of many antenna structures of various orders, it is observed that the value of R_{Gc} is important for obtaining an accurate circuit, but it is not significant in determining whether the inverting technique can be used. The reason why R_c and R_{Gc} are examined is because often times, the final poles used to generate the equivalent circuit fitted with an order between 2nd-N^{th}, often result in at least one complex pole pair despite the fact that the starting poles chosen are all real. For example, in many 4th order approximations with four real starting poles used, two complex pole pairs are produced during the fitting regardless of the iteration number. The proposed method of inverting resistances will be displayed through the use of applying it to the simulated design of a microstrip patch antenna at 2.4 GHz. This structure is shown in Figure 59. The $\lambda/2$ patch has a length, L, of 950 mils and a width, W, of 1050 mils. The substrate is a 76 mil thick layer of LTCC (ε_r = 5.4, tan δ = 0.0015). The S-parameters are obtained through simulation using MicroStripes 7.0. After transferring the S-parameters to admittance

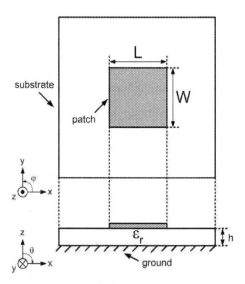

Figure 59. Illustration of simple microstrip antenna.

parameters, the data is fitted to a 4th order approximation using the following starting poles: -2.2e7 ± j*2.2e9 and -2.6e7 ± j*2.6e9. Lower orders of approximation of the admittance data produce inaccurate results and higher orders are unnecessary. Three iterations are carried out in this fitting process. Figure 60 shows a comparison between the acquired admittance data, y(s), and the fitted function $y_{fit}(s)$ versus frequency. A good agreement is observed in this plot with a deviation less than a factor of 10^{-4}. The 4th order approximation produced two complex pole pair branches as well as a resistor branch (R_0 = 8.35 kΩ) and a capacitor branch (C_0 = 174.15 fF). Table 3 shows the passive component values of the complex pole pair branches.

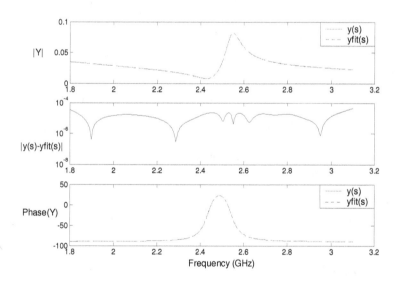

Figure 60. Admittance parameter comparison between y(s) and $y_{fit}(s)$ displaying the magnitude (1st row), magnitude deviation (2nd row), and phase (3rd row).

Table 3. Passive component values of design at 2.4 GHz.

4th order	CPP Branch 1	CPP Branch 2
R_c (Ω)	0.19	-6.11
L_c (H)	2.54e-9	3.04e-8
R_{Gc} (Ω)	-74.92	-32597
C_c (F)	2.43e-11	1.29e-13

In this table, the existence of negative resistor values is present. Consider the fact that the antenna being modeled has a dominant TM_{10} mode. There should be no parasitic resonances that exist in this design. Therefore, a new technique is suggested of inverting the negative resistance values to positive resistance values, hence, obtaining a physical equivalent circuit. Through experimentation, it has been observed that this technique works exceptionally well with conventional microstrip antennas with dominant TM_{10} resonant modes by accurately predicting the resonant frequencies/bandwidths as well as the value ranges of the S_{11} parameters. Agilent ADS2005 is used to simulate the equivalent circuit. A comparison between the S-parameters of the positive resistance circuit and $y_{fit}(s)$ versus frequency is shown in Figure 61. The positive resistant circuit agrees well with $y_{fit}(s)$. The magnitude deviation is on the order of 10^{-2}.

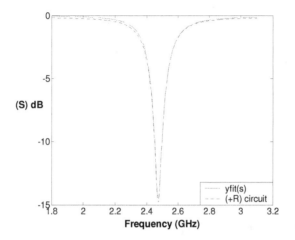

Figure 61. S-parameter plot comparing the fitted function, $y_{fit}(s)$, and the positive
resistance equivalent circuit of patch antenna at 2.4 GHz.

A second equivalent circuit has been derived for a microstrip antenna operating
at 66 GHz. The schematic for this structure is the same as that shown in Figure 59. The
length and width of the patch are 1350 μm and 1550 μm, respectively. The substrate
(RT/Duroid 5880: ε_r = 2.2, tan δ = 0.0009) has a thickness of 254 μm. A 3rd order
approximation with two iterations is used to fit the admittance data acquired through
simulation using the following starting poles: -68e7 ± j*68e9 and -73e9. When using a
3rd order approximation, at least one real pole has to be selected as a starting pole because
the complex poles must come in pairs. One real pole branch, one complex pole pair
branch, a resistor branch (R_0 = 8.35 kΩ) and a capacitor branch (C_0 = 174.15 fF) are
produced by this fitting. Table 4 shows the passive component values of the real pole
branch and the complex pole pair branch.

Table 4. Component values of design at 66 GHz.

3rd order	CPP Branch
R_c (Ω)	-33.02
L_c (H)	1.20e-9
R_{Gc} (Ω)	-12562
C_c (F)	4.71e-15
	RP Branch
R_r (Ω)	0.72
L_r (H)	1.72e-10

After inverting the resistors, an equivalent circuit is simulated. Figure 62 shows the return loss of the positive resistance circuit and y_{fit}(s) versus frequency. A good agreement is observed with a deviation on the order of milli-dBs.

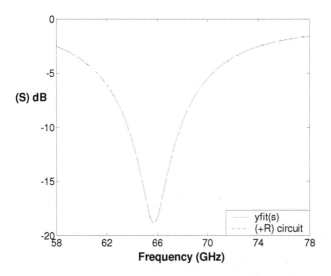

Figure 62. S-parameter plot comparing y_{fit}(s) and the positive resistance equivalent circuit of a patch antenna at 66 GHz.

In an attempt to develop a methodology for producing a physical circuit from negative resistances, the author experimented with various antenna designs, approximation orders, iterations, and other factors. From that experimentation, it is noticed that a strong correlation exists between low order approximations, first resonant mode structures, and the value of R_c. The significance of using an order number less than four to approximate simple antenna designs has been discussed, but it is observed that the magnitude of negative values of R_c must be small (less than 100 Ω) for the resistances to be inverted. In this above benchmarking case, the sole negative R_c value (= -6.11 Ω) is in this range; therefore, all resistances can be inverted. Despite the fact that one of the negative values of R_{Gc} is quite large, the author's investigation showed that as long as the values of R_c maintain a small value (< 10^2 Ω), the equivalent circuit would still be accurate. In addition, extensive numerical experiments demonstrated that after negative R_c and R_{Gc} values have been inverted, one may still use positive resistance values within ±5% of the inverted resistances to produce an equivalent circuit with insignificant variation in the S-parameter results.

6.3 Applications

In this section, the resistance invertibility technique has been applied to a circular loop resonant antenna and a frequency agile microstrip antenna with tuning capacitors. The simulated data for the fitting process is produced using MicroStripes 7.0. The equivalent circuits are simulated using Agilent ADS2005.

6.3.1 Circular loop antenna at 100 MHz

In this design, the author wanted to display the capability of producing an equivalent circuit from a circuit simulation instead of an electromagnetic simulation. A circular loop resonant antenna was designed to operate at 100 MHz. The data acquired from this antenna was obtained through a circuit simulator called RFSim. A picture of the circular loop is presented in Figure 63, and its circuit representation is shown in Figure 64. The value of the parallel inductance was obtained from connecting the circular loop to a coaxial connector and measuring the input reactance on a network analyzer. The parallel capacitance and resistance values were obtained through knowledge of the desired resonant frequency and quality factor. A series capacitor is used to tune the circuit to around 100 MHz. A 2nd order approximation with two iterations is used to fit this design. The two starting poles used in this fitting are -100e4 ± j*100e6.

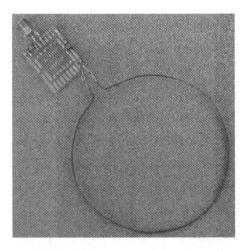

Figure 63. Illustration of a circular loop antenna at 100 MHz.

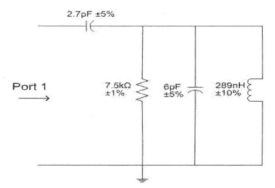

Figure 64. Circuit representation for a circular loop antenna at 100 MHz.

One complex pole pair branch, a resistor branch (R_0 = 63.49 kΩ) and a capacitor branch (C_0=2.35 pF) are produced by this fitting. The component values of the complex pole pair branch are R_c = -24.17 Ω, L_c = 2.88 µH, R_{Gc} = 86.16 kΩ, and C_c = 872.81 fF. After inverting R_c, these component values are placed in an equivalent circuit representation (shown in Figure 58) and simulated. Figure 65 displays the return loss of the positive resistance circuit in comparison to $y_{fit}(s)$ versus frequency. This figure demonstrates that the two plots are in agreement with each other except around the resonant frequency (100.4 MHz). Further investigation shows that the magnitude of the admittance of the plots around 100 MHz are different by a factor of 1/3. 3rd order through 10th order approximations were carried out with numerous iterations in an effort to match the responses, but obtaining a good agreement was unsuccessful.

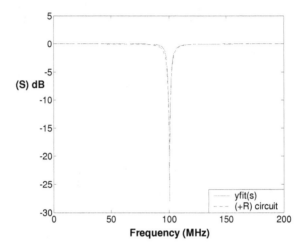

Figure 65. S-parameter plot comparing $y_{fit}(s)$ and the positive resistance equivalent circuit of a circular loop at 100 MHz.

6.3.2 Microstrip antenna with tuning capacitors (dual resonant mode fitting)

A microstrip antenna with tuning capacitors [62] is simulated with two resonant frequencies around 2.4 GHz. The schematic of this antenna is shown in Figure 66. By feeding the antenna along the diagonal of the patch, two resonant modes (TM_{10} and TM_{01}) are excited. The dimensions of the antenna are the same as that of the antenna in Figure 59. The value of both capacitors is 280 fF. After the admittance data is acquired, an 8th order approximation is applied using three iterations. The starting poles for this fitting are $-2.2e7 \pm j*2.2e9$, $-2.3e7 \pm j*2.3e9$, $-2.4e7 \pm j*2.4e9$, and $-2.5e7\pm j*2.5e9$. This approximation produced four complex pole pairs, a resistive branch ($R_0 = 180.81$ kΩ), and a capacitor branch ($C_0 = 114.11$ fF). Table 5 shows the component values of the complex pole pair branches.

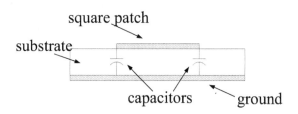

Figure 66. Illustration of a microstrip antenna with tuning capacitors.

Table 5. Component values of dual resonant mode design at 2.4 GHz.

8th order	CPP Branch 1	CPP Branch 2	CPP Branch 3	CPP Branch 4
R_c (Ω)	-0.03	-56.23	-5.98	481.35
L_c (H)	2.59e-9	1.13e-7	3.62e-8	2.00.e-7
R_{Gc} (Ω)	-322.92	1.15e6	-23905	-39783
C_c (F)	2.05e-11	4.07e-14	1.12e-13	1.01e-14

The resistors in these branches are inverted, and the circuit is simulated. The return loss versus frequency for the inverted circuit and $y_{fit}(s)$ is displayed in Figure 67. There is a disagreement in the return loss around 2.32 GHz, but this discrepancy is on the order of 10^{-2}. The author believes that the fitting for this structure is good because the dominant TM modes are created from feeding the patch directly. In future applications of this technique, the author plans to discuss how the results could have additional disagreement when the dominant TM modes do not get excited due to a direct feeding of the patch.

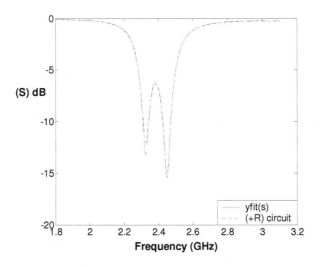

Figure 67. S-parameter plot comparing $y_{fit}(s)$ and the positive resistance equivalent circuit of a 2.4 GHz patch antenna with tuning capacitors.

6.4 Significance of Approximation Order and Number of Iterations

When an equivalent circuit is created, the number of passive elements used is dependent on the order of the approximation; hence, it is necessary to discuss the significance of this order. Upon performing research on many antenna designs, including the ones presented in this paper, it is concluded that there is a tradeoff that deals with the order number that is selected. If the order number is too small, it may be difficult to fit the peaks that exist in the admittance data. On the other hand, if the order is too large, more branches are created which can lead to more negative resistances that are too large to be inverted. Of course, the number of poles will depend on the order number (N poles $= N^{th}$ order). As stated earlier, the choice of starting poles is arbitrary because the type of poles may change in the fitting process. For example, four real starting poles selected for a 4th order approximation may produce two complex pole pairs or two real poles and one complex pole pair. There are situations where the chosen order number is much larger (10th or higher) than necessary for a simple antenna (2nd through 4th). After the fitting is completed, some of the branches may be physical with resistances small enough to be inverted, while other branches contain elements that lead to virtual open circuits (resistors $\sim 10^{12}$, inductors $\sim 10^{1}$, and capacitors $\sim 10^{-22}$). Accordingly, these branches can be removed and a lower order equivalent circuit will result. The significance on the number of iterations used in the fitting process is small if the magnitudes of the starting poles lie in the frequency range of the approximation. (If the approximation is between 1-2 GHz, for instance, then the magnitude of the starting poles must lie between 1-2 GHz.) The passive elements produced by the fitting tend to stabilize after the 2nd or 3rd iteration. The antenna in Figure 59 was fit using one through five iterations. One circuit was simulated for each number of iterations. The passive element values of the 3rd iteration are similar to those of the 4th and 5th iteration. Therefore, the circuits generated after the 3rd iteration had the best match to the approximated function with the smallest computational overhead.

CHAPTER 7

DEVELOPMENT OF LOW COST DUAL-FREQUENCY AND DUAL-POLARIZATION PATCH ANTENNA ARRAYS ON MULTILAYER LCP ORGANIC SUBSTRATES FOR SOP RF FRONT ENDS

Many radar and communication systems require antennas equipped with dual-polarization capabilities that give a higher capacity of data transfer. In multiple input, multiple output (MIMO) mobile communications systems, dual-polarized antennas serve as a means of increasing the number of sub-channels [63], while in automotive radar systems, dual-polarized antennas can be used to detect potential road hazards, such as black ice [64], that have a cross-section with one dimension being much thinner than the other in the perpendicular direction. In this case, these antennas can receive signals aligned on both axes contrary to the single polarizations utilized in the past. Moreover, dual-frequency antennas have gained interest in wireless communication systems where different frequency applications, including wireless local area networks (WLANs-802.11a, b, g) and personal communication services (PCS) can be covered in a single design. Over the last thirty years, antenna arrays have been utilized in various applications due to their directive main beam and high gain characteristics for long range communication.

When designing dual-frequency, dual-polarized microstrip antenna arrays in a multilayer environment, there are many factors that need to be considered. If the antennas are excited on different layers, the far-field radiation pattern of the lower-layer array may develop nulls in the main beam due to blockage from the array on the upper layer. Additionally, to achieve a ground plane size that is 2λ - 3λ (at the lower frequency), the same ground plane might be 7λ - 8λ at the higher frequency causing edge diffraction

122

that has the potential of contaminating the higher frequency far-field radiation pattern [65]. To realize a design that can be easily integrated with 3D modules containing MMICs, filters, and embedded passives, the antenna structure has to be thin with printed feedlines to simplify fabrication processes. Therefore, there is a need for a complex feeding structure that maintains the following properties: low interconnect losses, high impedance lines to minimize feedline radiation, and careful positioning of the feedlines to minimize cross-coupling of signals [38]. Cross-polarization must be minimal to reduce high sidelobe levels. Last, but not least, the distance of the antenna elements in the array has to be taken into account. A close spacing may limit the desired -3 dB beamwidth and directivity, while far spacing of antenna elements will generate sidelobes in the far-field radiation pattern when the center-to-center spacing between elements becomes greater than λ. Once these design considerations can be effectively addressed, it can be seen that multilayer vertical integration has many advantages over single layer antenna structures. By utilizing aperture-coupled and/or proximity-coupled feeding, feedline radiation is reduced when the feedlines are placed on a layer that is vertically close to the ground plane. This method of feeding can also minimize cross-polarization. Additionally, vertical integration capabilities allow the size of the lateral structure to be greatly reduced, allowing for compact implementations. In array design, expansion into larger arrays at each band is feasible and good isolation between the arrays can be achieved. Without vertical feeding, it would be quite difficult to feed two separate antenna arrays of distinct frequencies on the same layer. Consider two 2x2 antenna arrays of two different frequencies. One array could be fed with minimal complexity and degradation, but the other array would have to be excited by feedlines placed on the perimeter of the effective aperture area of the array to minimize cross-coupling between the feedlines. This could greatly increase the size of the structure as well as complicate the design's extension to a 4x4 or 16x16 element antenna array. One design of a dual-frequency, dual-polarized microstrip antenna array incorporating vertical integration is proposed by Granholm and Skou for the purpose of extracting multi-frequency synthetic aperture radar (SAR) data [66]. This design consists of L-band and C-band patches operating at around 1.25 and 5.3 GHz, respectively, on the metal layers separated by substrate layers of three distinct dielectric media including foam. Low cross-polarization has been reported for this probe-fed structure.

A promising alternative to mature expensive multilayer substrates, like LTCC, is LCP. This material has gained much consideration as a potential high performance microwave substrate and packaging material [14]. Its low dielectric constant ($\varepsilon_r = 3.0$) and low-loss performance (tan $\delta = 0.002$-0.004 for f $<$ 35 GHz) [67-69] is a key feature in minimizing dielectric and surface wave losses. Moreover, the near hermetic nature of the material (water absorption $<$ 0.04%) [70], the flexibility, and the relatively low processing temperatures enable the design of conformal antenna arrays, while the integration of RF MEMS devices and the low deployment costs in space applications from rolling antennas on LCP is quite attractive. The low water absorption of LCP makes the material stable in a variety of environmental conditions, hence preventing changes in the material's dielectric constant and loss tangent. The multilayer circuitry can be easily realized due to the use of two types of LCP substrates that have different melting temperatures. The high melting temperature LCP (around 315 °C) is primarily used as the core substrate layer, while the low melting temperature LCP (around 290 °C) is used as a bonding layer. The thickness of readily available LCP substrate layers vary between 1 and 4 mils, and this variance can be proven to be a significant advantage in complex 3D structures that require more flexibility in designing the total substrate thickness which can better meet strict design requirements. It is known that structures in LTCC can be made more compact due to its high dielectric constant and designing compact dual-polarization arrays on LCP can be a real challenge. However, the low dielectric constant of LCP will result in wider bandwidths and increased gain and efficiencies in comparison to LTCC materials due to the larger physical areas. Furthermore, the low cost of LCP (~$5/sq. ft. for 2 mil, single clad, low melt LCP) [71], 2-3 times less than LTCC, and its multilayer lamination capability makes it appealing for high frequency designs where excellent performance is required for a minimal cost.

This chapter presents, for the first time, the design and measurement of two dual-frequency (14 GHz and 35 GHz), microstrip antenna arrays with dual-polarization capabilities excited separately at each frequency on LCP multilayer substrates for an SOP RF front-end. These designs can be applied to the remote sensing of precipitation at 14 and 35 GHz. In addition, these designs can be extended by integrating RF MEMS switches with the antenna arrays to switch polarizations introducing the possibility of a low-power reconfigurable antenna array design.

7.1 2x1 Antenna Array Architecture

The generic multilayer architecture of the dual-polarization, dual-frequency 2x1 microstrip antenna array at 14 and 35 GHz is shown in Figure 68. The metal, used in simulation and fabrication for the ground plane and the antenna elements, is copper (Cu) that has a thickness of 18 μm. The total substrate thickness (h) for the design is 17 mils,

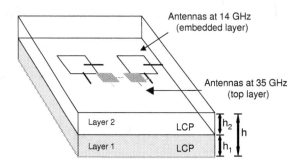

Figure 68. 2x1 antenna array architecture.

consisting of two LCP layers (each 8 mils thick) and a 1 mil bonding layer. The 35 GHz

antenna array is placed on the top surface of the LCP substrate (at the interface of LCP

and air), while the 14 GHz antenna array is embedded on an 8 mil layer (h_1) for

compactness, crosstalk minimization, and less radiation blockage. The two arrays (one

for each frequency) are designed independently and then, fine-tuned before integration to

optimize the impedance matching and radiation characteristics across both bands. The

feeding network for each band is placed on the same layer as the corresponding radiating

element to simplify the design and minimize the vertical cross-coupling between the two

single-frequency arrays. The feeding network of the embedded antenna can be measured

by either fabricating a via from the feedline to the top layer of the substrate and probing

the metal pad that is on top of the via or extending the feedline of the embedded antenna

to a point where the top laminated layer of the substrate no longer covers the feedline and

it is visible to be probed by a coaxial connection. Careful consideration must be taken to

minimize the coupling between the feeding network and the radiating element since both

of them are placed on the same layer. Despite the fact that a wider bandwidth can be achieved by aperture-coupling, it is stressed that bandwidth is not a critical requirement for these designs.

7.2 Antenna Structure and Analysis

The antenna structure for the proposed dual-frequency array is shown in Figure 69. The square patches are inset-fed for both the x- and y-directed feeds to match the input impedance of the antenna and achieve the maximum transfer of energy. The antennas are aligned parallel to the x- and y-axis at a center-to-center electrical distance of 0.93 λ_g for the 14 GHz design and 0.94 λ_g for the 35 GHz design. The feed network consists of a combination of T-junctions and quarter-wave transformers [72] that are used to match the antenna array. A 50 Ω feedline is used at the input terminal for the purpose of matching the 50 Ω coaxial probe utilized for the measurements. The feeding structure contains 200 μm gaps to simulate an "OFF" state which prevents the excitation of the dominant TM mode in one direction so that the dominant TM mode in the orthogonal direction can propagate. Based on the positions of the gaps and the design of the feeding network, the line lengths transferring the RF signal to each polarization for the 14 GHz array are equal. Therefore, the performance of the 14 GHz array should be similar for both polarizations. Differences in performance characteristics may stem from an additional discontinuity in the feeding of the y-directed polarization as well as the coupling of energy from the radiating edges of both polarizations to the radiation of the feedlines. For the 35 GHz design, the x-directed feed should perform slightly different than the y-directed feed due to the additional line lengths and discontinuities in the network. In an effort to minimize simulation time and maintain a similar phase for both antennas, an asymmetric T-junction was used in this design. The lengths of the arms of the T-junction were equal to each

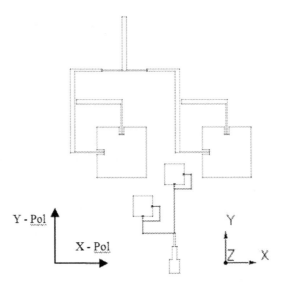

Figure 69. Antenna structure of 2x1 antenna array.

other despite the fact that one of the T-junction feedlines contains a discontinuity (bend). Although the phase mismatch of the antennas should be quite small or nonexistent, these bends can produce undesired feedline radiation [73], and as a result, parasitic modes can be excited. Careful attention is placed on the distance of the feedlines from the antennas and other feedlines to minimize crosstalk and cross-coupling of energy which could excite parasitic modes. All the metals of the design (14GHz and 35 GHz array) are oriented to the x- and y-axis. Each array is designed separately, then integrated together to complete the architecture. The center of the antenna in the 35 GHz structure that is closest to the 14 GHz array is positioned equidistant from the radiating edges of the x-directed feed with respect to the x-axis. This position is determined by the desire to minimize blockage effects in the radiation pattern while maintaining a compact structure suitable for expansion to a 2x2 design. The antenna arrays are simulated using MicroStripes 6.0, allowing for finite-size ground planes to be realized, so backside radiation and edge diffraction effects are accounted for in the simulation. In addition, this time domain simulator handles structures aligned on the x-y axis pretty well due to the meshing of the Cartesian grid.

7.3 Fabrication and Measurement Setup

Both 14 and 35 GHz antenna arrays were fabricated with double copper-clad 8 mil LCP dielectric sheets from Rogers Corporation. A standard photolithographic process was used for fabrication. Shipley 1827 photoresist was used for pattern definition. Both arrays were then exposed under 16,000 dpi mask transparencies pressed into sample contact with 5" glass mask plates. Then, photoresist development and a wet chemical etch with ferric chloride were performed to complete the antenna patterning. During the copper etch for the 14 GHz array, a protective tape was used to stop the bottom ground plane from being etched. The bottom metallization layer of the 35 GHz array was purposely etched during the patterning in preparation for the concurrent bonding to the LCP substrate on which the 14 GHz array was patterned. The LCP layers with the 14 GHz and the 35 GHz array were then bonded together in a Karl Suss SB-6 silicon wafer bonder using a 1 mil low melt LCP bond layer sandwiched between the two 8 mil high melt LCP core layers. The microstrip feed line extended beyond the upper substrate for a sufficient length to allow for a coaxial connection without the need of a viahole. Illustrations of the fabricated structures and one previous 2x1 antenna array showing the flexibility of the LCP substrate are displayed in Figure 70.

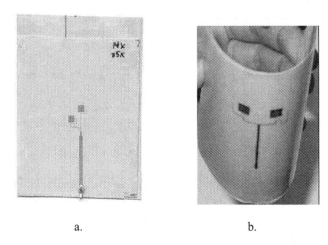

a.   b.

Figure 70. a.) Fabricated 14 GHz and 35 GHz array and b.) a flexibility demonstration.

The arrays were mounted on an aluminum fixture that included a coaxial-to-microstrip connector to facilitate the S-parameter measurement. A short, open, load, and thru (SOLT) [74] calibration was performed with the reference planes at the end of the coaxial cables. The antennas-under-test (AUT) were mounted on an aluminum block with a microstrip to 2.4 mm coaxial launcher. Before radiation pattern measurements were performed, the resonant frequency of the AUT was measured on a vector network analyzer, and when required, the microstrip launcher was adjusted to improve the AUT to coaxial launcher impedance match. An anechoic chamber with the AUT functioning as the transmitting element and a 15 dB gain horn antenna functioning as the receive antenna was used for radiation pattern measurements. There was 1 m spacing between the two antennas, which corresponds to three and five times the far field limit at 14 and 35 GHz, respectively. The diode detector was calibrated and the transmitter power was adjusted to maximize detector sensitivity while operating in its linear region. The AUT was rotated through the measurement plane, and the entire system, including the data recording, was automated. Because the microstrip launcher and the absorbing material placed around it covered a portion of the plane during the scan, there was a slight asymmetry in the radiation patterns due to the characterization system. In addition, the absorber affected the radiation pattern at scan angles greater than 70° off boresight.

7.4 2x1 Simulation and Measurement Results

The simulated and measured return loss plots versus frequency for the 14 and 35 GHz array are shown in Figure 71. Only the x-directed feed for this configuration is considered. The dual-frequency array is excited at one frequency, while the other array is treated as a parasitic element. The results are summarized in Tables 6 and 7.

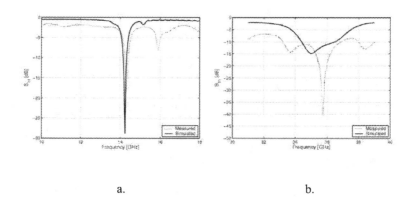

a.   b.

Figure 71. Return loss versus frequency of the a.) 14 GHz and b.) 35 GHz array.

Table 6. Return loss characteristics of the 2x1 Array at 14 GHz.

Characteristic	Simulated 14X	Measured 14X
Resonant Frequency	14.237 GHz	14.32 GHz
Return Loss (S_{11})	-28.8 dB	-18.2 dB
-10 dB Return Loss Bandwidth	170 MHz	165 MHz
Percent Bandwidth	1.17%	1.15%

Table 7. Return loss characteristics of the 2x1 Array at 35 GHz.

Characteristic	Simulated 35X	Measured 35X
Resonant Frequency	35.01 GHz	35.71 GHz
Return Loss (S_{11})	-15 dB	-40.3 dB
-10 dB Return Loss Bandwidth	2060 MHz	3360 MHz
Percent Bandwidth	5.88%	9.41%

The measured return loss for the 14 GHz array exhibits a frequency shift (to the right) of 320 MHz. There is a parasitic resonance around 16 GHz caused by undesired radiation and cross-coupling of energy due to the lack of symmetry. The impedance bandwidths for the simulated and measured return loss plots are in good agreement. The measured return loss for the 35 GHz design is better than the simulated results. This difference can be attributed to the change in the inset depth of the fabricated design. The 35 GHz array is first simulated and optimized separately before it is integrated into the dual-frequency design. Then, the simulated dual-frequency structure excited at 35 GHz is optimized before it is fabricated. The return loss of the simulated individual 35 GHz array is lower than that of the simulated dual-frequency array excited at 35 GHz due to the parasitic presence of the underlying 14 GHz array. Taking advantage of the etching process in the fabrication, the inset depth was altered (about 13-15 um) in the fabricated design. The inset depth of the fabricated design provided a better match to the feedline than the inset depth used in simulation. If the inset depths were the same for the simulated and fabricated designs, a closer agreement in the return loss could be expected. There is a frequency shift (to the right) of 710 MHz in the measured plot which can be attributed to fabrication and measurement tolerances. The impedance bandwidth in the measured return loss is significantly greater than that of the simulated plot, but a parasitic mode exists in the plot that is close to the TM_{10} resonance, which is the major factor in the acquisition of a wide measured bandwidth. This parasitic resonance could have been caused by feedline radiation, measurement inaccuracies, line losses at transitions, or connector losses. Without this parasitic resonance, the measured impedance bandwidth of the TM_{10} mode is expected to be similar to the simulated bandwidth.

Additionally, the simulated and measured 2D radiation patterns are shown in Figure 72 (for the E- and H-plane at 14 GHz) and Figure 73 (for the E- and H-plane at 35 GHz). The results are summarized in Tables 8 and 9. The E- and H-plane beamwidths for the measured patterns are lower than the simulated results for the 14 GHz array. Perhaps, some of the electromagnetic fields from the feed network have coupled power to the antennas causing a greater effective aperture and hence, a reduced beamwidth in the measured response. It is also observed that the H-plane beamwidth is wider than the E-plane beamwidth; this is a result of the design configuration for the 14 GHz array. Furthermore, since the 14 GHz array is excited in the x-direction and the antennas in the array are along the x-direction, the array factor is observed in the E-plane (not the H-plane). Therefore, an x-directed feed will have a greater H-plane beamwidth than E-plane beamwidth. Maximum cross-polarization levels for the simulated and measured patterns are comparable. The measured co-polarized components for the E- and H-plane

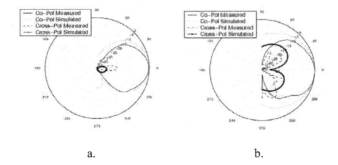

a.   b.

Figure 72. a.) E-plane and b.) H-plane radiation patterns at 14 GHz.

Table 8. Radiation pattern characteristics of the 2x1 Array at 14 GHz.

Characteristic	Simulated 14X	Measured 14X
E-Plane -3 dB Beamwidth	48^0	37°
H-Plane -3 dB Beamwidth	77^0	55°
Max. Cross-pol. Level (E-plane)	-32 dB	-25 dB
Max. Cross-pol. Level (H-plane)	-16 dB	-16 dB

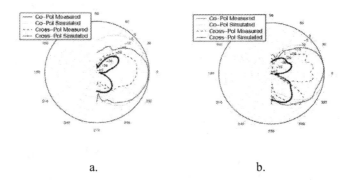

a.   b.

Figure 73. a.) E-plane and b.) H-plane radiation patterns at 35 GHz.

Table 9. Radiation pattern characteristics of the 2x1 Array at 35 GHz.

Characteristic	Simulated 35X	Measured 35X
E-Plane -3 dB Beamwidth	69^0	48°
H-Plane -3 dB Beamwidth	59^0	45°
Max. Cross-pol. Level (E-plane)	-21 dB	-9 dB
Max. Cross-pol. Level (H-plane)	-15 dB	-8 dB

of the 35 GHz array are quite similar due to the symmetry of the antennas in the array (not the feed network). The simulated E-plane beamwidth is wider than the measured E-plane beamwidth because of the blockage from the absorber that was placed above the long 50 Ω feedline to prevent its radiation. Furthermore, the E-plane beamwidth of the 35 GHz array is slightly larger than the H-plane beamwidth due to the center-to-center spacing of the antenna elements. In addition, the measured patterns show significant unwanted levels of cross-polarization. This is primarily due to the complex feeding structure and the coupling of feedline radiation from lines that are vertically close (200 µm) to each other. In addition, it has been noted in [11] and [75] that the cross-polarization level tends to increase as the substrate thickness increases. Therefore, the higher frequency (35 GHz) antenna array on the electrically thicker substrate could exhibit a worse cross-polarization level than the lower frequency (14 GHz) array on the electrically thinner substrate. An improved design would place the 35 GHz patches on a thinner (e.g. 100 µm) LCP substrate. It should also be noted that the previous results were repeatable after the antennas were flexed for several times and additional measurements were taken.

7.5 2x2 Antenna Array Architecture and Structure

The generic multilayer architecture of the dual-polarization, dual-frequency 2x2 microstrip antenna array at 14 and 35 GHz is shown in Figure 74.

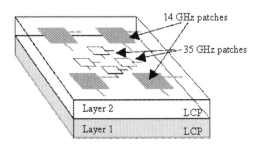

Figure 74. 2x2 antenna array architecture.

This antenna is designed on a similar LCP substrate with an 18 μm metal (Cu) thickness. The total substrate thickness (h) for the design is 17 mils, which is comprised of two LCP layers (each 8 mils thick) and a 1 mil bonding layer. The 35 GHz antenna array is placed on the top surface of the LCP substrate (at the interface of LCP and air), while the 14 GHz antenna array was embedded on an 8 mil layer (h_1). Again if one considers the placement of the 35 GHz array on an electrically thicker substrate ($0.081\lambda_g$) than the placement of the 14 GHz array ($0.016\lambda_g$), it becomes important to optimize the design of the feeding network of the 35 GHz design to minimize blockage effects that could degrade the radiation pattern at 14 GHz. The antenna structure for the proposed dual-frequency array is shown in Figure 75.

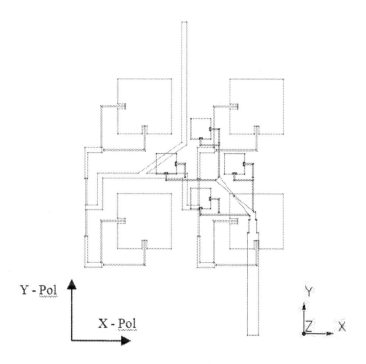

Figure 75. Antenna structure of 2x2 antenna array.

The position of the 35 GHz array is carefully aligned to exist inside the center area of the 14 GHz array to create horizontal, vertical, and diagonal symmetry between the antenna elements (without the feeding network). This design utilizes a combination of T-junctions and quarter-wave transformers in order to feed both of the arrays. In the feeding network of the 14 GHz array, the position is selected in order to maintain equal phasing of the individual elements so a broadside radiation pattern could be obtained. Despite slight changes in the impedances of the microstrip lines, the framework of the 2x1, 14 GHz array is used as a model for the 2x2 array. Similarly, the framework of the 2x1 configuration is used to construct the 2x2, 35 GHz array. The feeding network of the

2x2, 35 GHz array is more complex because it was necessary to maintain the placement of the antennas with respect to a Cartesian grid to minimize the computational time when using a Cartesian grid-like simulator. Due to this placement, it was difficult to determine a position to feed the array where all four elements could maintain equal phases. To overcome this shortcoming, the 2x2, 35 GHz array makes use of two impedance transforming exponentially-tapered transmission-lines; each one has an electrical length of $\lambda/2$. Therefore, the total length is λ, and the additional two elements are in phase with the original two elements in the 2x1 array. Similarly in this design, the feeding structure contains 200 μm gaps to simulate an "OFF" state which prevents the excitation of the

dominant TM mode in one direction so that the dominant TM mode in the orthogonal direction can propagate. Each array was designed separately, then integrated together to complete the architecture. Unfortunately, since the size of the array is larger and small discretized cells ($< \lambda/100$) are needed to accurately model the higher frequency array (35 GHz), the simulation package MicroStripes 6.0 is unable to simulate the integrated arrays due to insufficient memory. Additional attempts to simulate this design using three additional commercial simulation packages were performed, but again, the same problem of insufficient memory was observed. Based on approval from the project leaders on the optimization of the individual 2x2 arrays, the decision was made to skip the simulation process of the integrated design and proceed to the next steps of fabrication and measurement. The setup for the fabrication and measurement for this design is similar to that of the 2x1 arrays, and therefore, its details will not be repeated.

7.6 2x2 Simulation and Measurement Results

The simulated and measured return loss plots versus frequency for the 14 and 35 GHz array are shown in Figure 76. Only the y-directed feed for this configuration is considered. The dual-frequency array is excited at one frequency, while the other array is treated as a parasitic element. The results are summarized in Table 10. The measured return loss for the 14 GHz array exhibits a frequency shift (to the right) of 140 MHz. This value represents a shift of 1%. The return loss at the TM_{10} resonance is quite low; hence, a well matched antenna is obtained. The measured bandwidth is slightly larger for the 2x2 array at 14 GHz in comparison to the 2x1 design because additional loss of energy

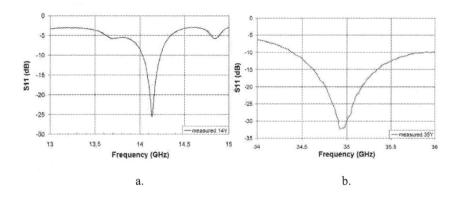

a.   b.

Figure 76. Return loss versus frequency of the a.) 14 GHz and b.) 35 GHz array.

Table 10. Return loss characteristics of the 2x1 Array at 14 and 35 GHz.

Characteristic	Measured 14Y	Measured 35Y
Resonant Frequency	14.14 GHz	34.92 GHz
Return Loss (S_{11})	-25.6 dB	-32.2 dB
-10 dB Return Loss Bandwidth	200 MHz	1580 MHz
Percent Bandwidth	1.41%	4.52%

between 13.6-13.8 GHz is exhibited due to a parasitic resonance at around 13.7 GHz. The measured return loss for the 2x2 array at 35 GHz also shows a frequency shift, but of a smaller value (80 MHz) than that of the 2x1 array at 35 GHz (720 MHz). The return loss at the TM_{10} resonance is quite low which shows that a well matched circuit is obtained. The most noticeable difference in the return loss plot of this array in comparison to the 2x1 array at 35 GHz is the decrease in the expected bandwidth. Although the measured bandwidth of the 2x1 design at 35 GHz is close to 10%, this value consists of the bandwidths of the TM_{10} resonance and a parasitic resonance that is close to the TM_{10} resonance. The expected bandwidth is around 5.5-6% based on the simulated results, but the bandwidth of the 2x2 array is around 25% smaller. This bandwidth would be smaller if the out-of-band energy loss was closer to 0 dB. The smaller bandwidth is due to spurious radiation of the feeding structure especially that of the two exponentially-tapered delay lines. Despite these effects, the profiles of the return loss plots are good.

Additionally, the measured 2D radiation patterns at 14 and 35 GHz are shown in Figure 77. The results are summarized in Table 11. The beamwidths and overall co-polarized radiation patterns for the E- and H-plane of the 14 GHz, 2x2 array are similar, which is expected due to the symmetry of the design. Additionally, the cross-polarization levels of both planes are around -20 dB again due to symmetry. These values are acceptable for this application. The co-polarized component of E-plane radiation pattern at 35 GHz exhibits a null at approximately 30°. The exact cause of the null is unknown to the author, but it is speculated that edge diffraction effects can be associated to this occurrence. Edge diffraction commonly occurs when the length and width of a finite

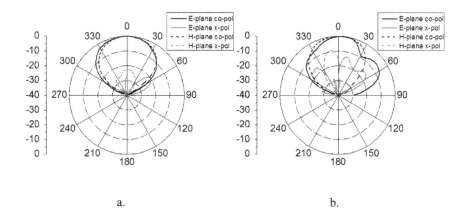

Figure 77. Radiation patterns of a.) 14 GHz and b.) 35 GHz antenna array.

Table 11. Radiation pattern characteristics of the 2x2 Array at 14 and 35 GHz.

Characteristic	Measured 14Y	Measured 35Y
E-Plane -3 dB Beamwidth	54^0	38°
H-Plane -3 dB Beamwidth	52^0	53°
Max. Cross-pol. Level (E-plane)	-20 dB	-10 dB
Max. Cross-pol. Level (H-plane)	-22 dB	-13 dB

ground is greater than 2-3 λ_0. At 35 GHz, the free space wavelength is 8.6 mm. The length and width of the ground plane is 45 mm (5.3 λ_0) and 50 mm (5.8 λ_0), respectively. These values are large enough to show that edge diffraction could play a role in the E-plane radiation pattern. It is interesting to note that a null does not appear in the co-polarized component of the H-plane radiation pattern at 35 GHz. Perhaps, the 14 GHz array below blocks the energy from leaking to the edges of the substrate in the H-plane. Due to these effects and the null in the E-plane, the beamwidth is less than expected. The cross-polarization is quite high for the 35 GHz design. This can be attributed to the edge diffraction effects which can degrade the co-polarized pattern and increase the cross-

141

polarization. In addition, the 35 GHz array is mounted on an electrically thicker substrate than the 14 GHz array; therefore, the cross-polarization level is higher. A soft surface structure for the smaller 35 GHz array can be incorporated to improve the radiation pattern and lower the cross-polarization level at this frequency.

CHAPTER 8

HIGH GAIN MICROSTRIP YAGI ANTENNA ARRAY WITH A HIGH FRONT-TO-BACK RATIO FOR WLAN AND MILLIMETER-WAVE APPLICATIONS

Although most WLAN applications utilize omni-directional antennas, directional and quasi-endfire antennas, such as Yagi arrays, have been employed to suppress unwanted RF emissions as well as unwanted interference in other directions. Yagi arrays have been utilized in ISM applications at 2.4 GHz (where directional radiation is necessary for long distance wireless communications and point-to-point communications), high performance radio local area network (HIPERLAN) and IEEE802.11a (IEEE802.11n) WLAN bands between 5-6 GHz (where quasi-endfire radiation in the range between 0-90° enables efficient communications at rates between 20-54 megabits per second), and ultra-broadband, millimeter-wave applications (above 30 GHz) that require high gain (> 10 dBi), quasi-endfire radiation to alleviate propagation loss effects through line-of-sight reception of waves at angles off broadside for wireless video transfer, millimeter-wave ad-hoc sensor networks, and point-to-multipoint wideband links. The Yagi arrays currently on the market [76-77] that operate at the lower frequencies (under 10 GHz) are too bulky and are unsuitable for compact integration with MMICs and other RF circuitry due to the size of the array.

There have been many printed Yagi antenna configurations that have been presented over the last fifteen years [78-84]. The use of Yagi antenna designs in microstrip technology was first proposed by Huang in 1989 which consists of four patches (one reflector, one driven element, and two directors) that are electromagnetically coupled to one another to create a steering of the main beam to a peak between 30-40°

[78]. To increase the gain, Densmore and Huang introduced microstrip Yagi arrays (each consisting of four elements) in four rows, while exciting the driven elements simultaneously [79-80]. This design can achieve a peak gain as high as 14 dBi, but F/B ratios around 4-5 dB may not be suitable for some applications. In [81], a successful attempt to improve the gain of a single microstrip Yagi array using a periodic bandgap (PBG) structure was proposed. This approach takes away from the manufacturing simplicity of using a double-clad board. In general, there is a tradeoff between achieving a high gain and maintaining a high F/B ratio and low cross-polarization for existing printed Yagi antenna arrays. Additionally, there has been one Yagi array that has been proposed for millimeter wave frequencies that radiates in the endfire direction with a high gain [82].

A new microstrip Yagi array design is presented that can achieve a high gain (> 10 dBi) without compromising the low cross-polarization and high F/B ratio qualities that are essential for the design of planar antenna architectures that can be easily integrated with wireless communications devices. Note in this paper, the F/B ratio is the ratio of frontside radiation in the range of $0° \leq \theta \leq 90°$ to backside radiation in the range of $-90° \leq \theta \leq 0°$ in the E-plane.

8.1 Antenna Structure

The proposed antenna is shown in Figure 78. It consists of seven patch elements as well as the feeding structure. The reflector elements, R, are treated as one single element with a gap through the middle to simplify the analysis. The driven element, D, is excited by a simple feeding structure through a gap in the reflector elements. The feeding

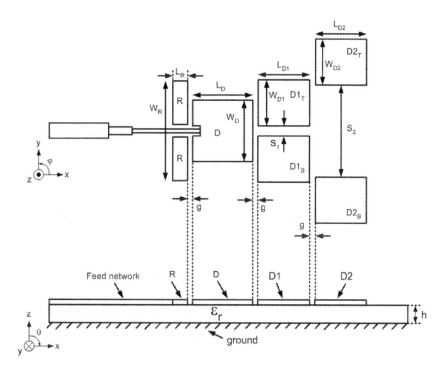

Figure 78. Proposed Yagi antenna structure.

structure consists of a 50 Ω feedline that is transformed to a high impedance line through the use of a quarter-wave transformer. High impedance lines (lines that have linewidths that are less than 0.15 times a patch length) are utilized to ensure that the feedline radiation close to the driven element will not disrupt the radiation of the antenna and lead to an increased F/B ratio. The remaining elements include four directors: $D1_T$ (top director1), $D1_B$ (bottom director1), $D2_T$ (top director2), and $D2_B$ (bottom director2). This antenna structure is designed on a double copper (Cu) clad board of RT/duroid 5880 material ($\varepsilon_r = 2.2$). The lengths and widths of the patch elements are denoted as follows: for the reflector, R, the length and width are L_R and W_R; for the driven element D, the length and width are L_D and W_D; for the director1 elements $D1_T$ and $D1_B$, L_{D1} and W_{D1} represent the length and width, respectively; and the length and width of the director2 elements $D2_T$ and $D2_B$ are L_{D2} and W_{D2}, respectively. The distance between the elements along the x-axis is denoted by g (note that these distances are the same). Furthermore, the distances between the director1 and director2 elements are represented by S_1 and S_2, respectively. The thickness of the substrate, shown in Figure 78, is denoted by h.

8.2 Principles of Operation

The mechanism of gain enhancement can be seen through the constructive interference between the R-D-D1$_T$-D2$_T$ and the R-D-D1$_B$-D2$_B$ single microstrip Yagi arrays. This constructive interference can be viewed as one that is analogous to the radiation mechanism of a patch antenna and its radiating slots. For the design proposed in Figure 78, the single R-D-D1$_T$-D2$_T$ array has a peak radiation at $10° \leq \varphi \leq 16°$, while the R-D-D1$_B$-D2$_B$ array radiates maximally at $-16° \leq \varphi \leq -10°$. The microstrip Yagi array, in Figure 78, has a maximum radiation at $\varphi = 0°$ with a higher gain. Consequently, the increased gain of the new antenna structure considerably decreases the backside radiation. This is due to the constructive interference between the single microstrip Yagi arrays that pulls more power away from the reflector side of the driven element and towards the $\varphi = 0°$ quasi-endfire direction. This decrease allows the effect of the reflector patch to be minimal. Through simulation, the length of the reflector, L_R, is designed to be about ¼ of the total width, W_R. A shorter length can result in increased backside radiation. A larger length is unnecessary for reducing the backside radiation (it will only increase the size of the design, while the radiation performance will stay the same). Feeding through the reflector has a negligible effect on the impedance mismatch between the microstrip-

CPW transition because the impedance difference between the lines is less than 5 Ω and both of the lines have impedances above 100 Ω.

The design of the microstrip Yagi array proposed in this paper consists of two major effects that are taking place to result in a higher gain and a higher F/B ratio; the first is the role of D1 elements (D1$_T$ and D1$_B$) and the spacing between them (S$_1$), while the second is the role of D2 elements (D2$_T$ and D2$_B$) and the spacing between them (S$_2$). The D1 elements are used to establish the directionality of the beam as well as increase the impedance bandwidth of the antenna due to the close proximity between the resonant modes of D1 and the driven element. (Note that since the resonant length of D1 is slightly shorter than the driven element, it will resonate at a slightly higher frequency, but the combination of these modes will produce an increase in the bandwidth.) On the other hand, the D2 patches are used to increase the gain of the design. Both concepts are explained further.

The S$_1$ parameter has to be small in order to strongly couple fields from the driven element to the D1 elements. When a parasitic element is placed next to a driven element, its distance has to be small (< 0.05 λ_{eff} where λ_{eff} = c/(f$_r$*ε_{eff}^0.5) and ε_{eff}, the effective dielectric constant, lies in the range of 1< ε_{eff} < ε_r) in order to increase the gap capacitance between the driven and D1 elements. An equivalent circuit that models the gap between the driven element and one D1 element is illustrated in Figure 79 where C$_F$ and C$_G$ are the fringe field and gap capacitances, respectively. Equations for solving for C$_F$ and C$_G$ are presented in [73]. As the D1 elements are separated further apart from each other, the gap capacitance decreases because the parallel plate area, A (A=t$_C$*W$_P$

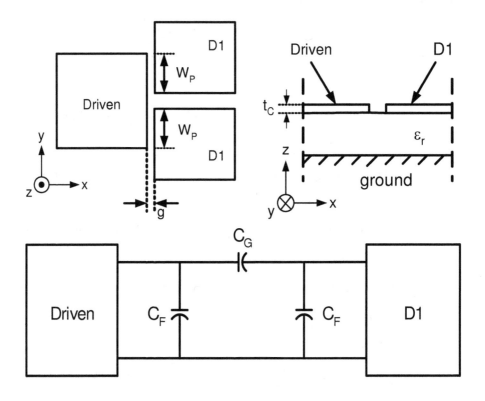

Figure 79. Schematic of driven element and D1 element and its equivalent circuit.

where t_C is the thickness of the conductor and W_P is the parallel plate width), between the elements is smaller (W_P decreases). This decrease in the gap capacitance results in a decrease of electric field coupling to the resonant D1 elements. The surface current distributions in Figure 80 display this effect.

 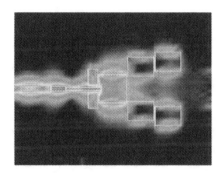

a.   b.

Figure 80. Surface current distributions on the conductors for
a.) $S_1 = 0.04\ \lambda_{eff}$ and b.) $S_1 = 0.40\ \lambda_{eff}$.

The D2 elements serve the sole purpose of enhancing the gain of the antenna. It is observed through simulation that a small value of S_1 (close placement of the D1 elements with respect to each other) has a major effect on the radiating edges of the D1 elements. In particular, the radiating edges of the D1 elements act as a single radiating edge with width 2W. Considering the strong magnitude of the electric fields at the edge as well as the strong fringing fields of the closely spaced D1 elements, the placement of the D2 elements is arbitrary with respect to receiving enough fringe field coupling to necessitate radiation. In other words, the D2 elements can be placed at many positions along the y-direction and still receive coupling (S_2 can range in value). Although, there are many possibilities of S_2 to receiving coupling, this parameter does have a significant effect on the gain of the antenna. As S_2 increases, the separation between the D2 elements also increases and the gain is improved because the effective aperture is electrically wider. The surface current distributions (Figure 81) show this effect.

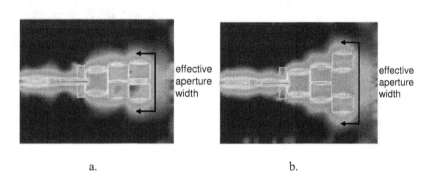

a.   b.

Figure 81. Surface current distributions on the conductors for
a.) $S_2 = 0.55 \lambda_{eff}$ and b.) $S_2 = 0.10 \lambda_{eff}$

There is a limit on the value of S_2. If S_2 becomes too large ($>0.55 \lambda_{eff}$), there will be insufficient coupling between the D1 and D2 elements and the enhancement of the gain will be suppressed.

Similar to what has been stated in previous microstrip Yagi papers [81], the addition of a 3rd set of director elements (D3) does not improve the gain of the Yagi array. This is due to the fact that the fields in the aperture become weaker as they propagate away from the D1 elements.

8.3 Initial Design

A design has been simulated using MicroStripes 6.5 as a proof of concept to cover a band above and below 32.5 GHz suitable for millimeter-wave applications. Figure 82 shows a plot of the return loss versus frequency for this design. The values of the design parameters are as follows: L_R = 1000 μm, W_R = 4016 μm, L_D = W_D = 2956 μm, L_{D1} = L_{D2} = 2814 μm, W_{D1} = W_{D2} = 2014 μm, S_1 = 286 μm, S_2 = 3686 μm, and g = 140 μm. The substrate thickness is 254 μm. These values represent the optimized values for this design in terms of highest achievable directivity and F/B ratio greater than or equal to 15 dB. There are two resonances that are present in the plot. The lower frequency

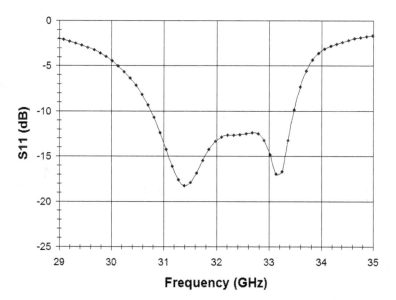

Figure 82. Return loss versus frequency of Yagi structure in Figure 78 covering band above and below 32.7 GHz.

resonance is the TM_{10} resonance of the driven element, D, while the higher frequency resonance is associated with the D1 elements that have a slightly shorter length. The optimized bandwidth is 8.3% due to the close proximity of the two resonances. The E-plane radiation patterns are shown at four frequencies (31.7, 32.3, 32.7, and 33.1 GHz) in Figure 83a, while Figure 83b shows the E-plane patterns at the -10 dB low and high edge frequencies (f_L and f_H). The scales are normalized with respect to the maximum E-plane radiation of the antenna. It is observed that as the frequency increases, the F/B ratio decreases. At 32.3 GHz, there is no significant backlobe present and the beamwidth is slightly wider. This can be attributed to the fields that are more concentrated around the driven element which resonates around that frequency.

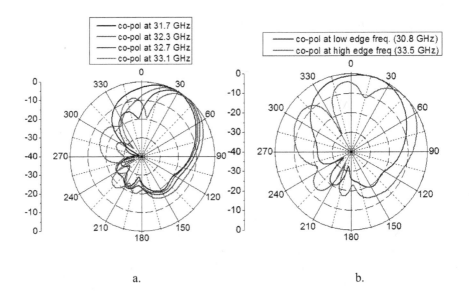

a.   b.

Figure 83. E-plane co-polarized components of Yagi structure in Figure 78 a.) at 31.7, 32.3, 32.7, and 33.1 GHz and b.) at 30.8, and 33.5 GHz.

The directivity is 11.1 dBi, and the scan angle is 31.5°. At 32.7 GHz, the F/B ratio is 15 dB and a scan angle of 35.5° with an E-plane (-3 dB) beamwidth of 40° is observed at this frequency. This F/B ratio is much higher than some values previously published that are between 5-10 dB [79-83]. A directivity of 11.4 dBi is achieved at this frequency. At 33.1 GHz, there exist two backside lobes that are -15 dB and -8 dB down from the main beam. In this case, the radiation from the D1 elements is stronger than the radiation from the driven element. Although there is a significant contribution of radiation from the driven element, the fields of the driven and D1 elements are out of phase. The directivity at this frequency is 11.1 dBi, and the scan angle has increased to 42°. At f_L, the pattern exhibits strong broadside characteristics, while at f_H, the grating lobes are significant in comparison to the main beam.

Although the impedance bandwidth is 8.3%, the usable radiation bandwidth is smaller, but this is dependent on the application and the desired specifications of F/B ratio and directivity. If a designer wants a 14 dB F/B ratio with a directivity greater than 10 dBi and a maximum radiation angle greater than 20°, then the radiation bandwidth may be less than 1%. However, (as stated in the following example) if a designer wants only a 10 dB F/B ratio, then the radiation bandwidth may be as much as 7%. The radiation pattern starts to exhibit the quasi-endfire radiation characteristic (seen in microstrip Yagi arrays) at approximately 31.5 GHz. At frequencies below 31.5 GHz, the antenna radiates a broadside wave due to the current distribution that exists mainly on the driven element. Between 31.5-32.7 GHz (bandwidth = 3.7%), a F/B ratio greater than 15 dB can be achieved. A greater than 40° beamwidth is observed between 31.5–33.2 GHz (bandwidth = 5.2%). In addition, limits on the gain and the angles of maximum radiation will help determine what the radiation bandwidth will be for a specific application. The impedance bandwidth and radiation bandwidth will be approximately equal when the acceptable F/B ratio is greater than 4 dB. The azimuth radiation patterns (on the plane parallel to the antenna) at the angle, θ, of maximum radiation are shown in Figure 84 at 31.7, 32.3, 32.7, and 33.1 GHz. The E_θ components of the azimuth plane radiation pattern illustrate a reduction of greater than 16 dB at φ = 180° with respect to the direction of maximum radiation along the positive x-axis (φ = 0°) for frequencies greater than 32.3 GHz. The azimuthal beamwidths are between 72-75°. Low levels of cross-polarization are obtained at all frequencies considered in this design.

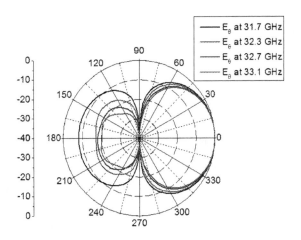

Figure 84. E_θ components at 31.7, 32.3, 32.7, and 33.1 GHz.

8.4 Parametric Analysis

There are many parameters that need to be addressed in this antenna structure due to the number of elements, coupling gaps, and distances between patches. This section gives insight on how the critical parameters affect the radiation characteristics; tables are presented in this section to show the comparisons. These parameters are varied around the optimized values given in the previous section. The simulated cross-polarization is better than 30 dB below the main beam and the -3dB beamwidths are between 40-42° in all of the parameters analyzed, so no mention of these characteristics will be discussed in this section.

8.4.1 Variation of width, S_2, at 32.7 GHz

The dimension, S_2, is the first parameter that was varied, and the affects of this variation are summarized in Table 12. For this variation, all variables are initially set to the optimized values given in Section 8.3. From this table, it can be observed that this parameter has a major affect on the directivity and F/B ratio because this parameter increases or decreases the effective aperture width of the antenna. When this value is larger, the directivity and F/B ratio is greatly increased. Smaller values have a severe effect on the F/B ratio. As the value of L_2 is increased, the antenna size in the y-direction is increased. A width of 3686 μm (~ 0.55 λ_{eff}) represents the largest possible value in which enough energy is coupled from the D1 elements. If S_2 becomes larger, energy from D1 will not couple to D2 (as explained earlier). The highest directivity and F/B ratio that can be obtained for this design is achieved at this value.

Table 12. Performance effects upon variation of width, S_2.

S_2 (μm)	Directivity (dBi)	Main beam tilt angle (deg)	F/B Ratio (dB)
3686	11.4	35.5	15
3086	11.3	38	10
2486	11.3	36.6	8
1886	11.1	36.6	8
1286	10.8	37	6
686	10.6	36.7	5

8.4.2 Variation of width, S_1, at 32.7 GHz

The variation of width, S_1, is summarized in Table 13. Again, all variables are initially set to the optimized values discussed in the last section. From the table, it is seen that the directivity and the F/B ratio is critically affected by the change in S_1. As S_1 increases, the directivity is decreased by as much as 2.4 dBi, while the F/B ratio is reduced by as much as 13 dB. The major reason for this decline in directivity and F/B ratio is the decreased coupling from the driven element to the D1 element as a result of the decreased gap capacitance. The main beam tilt angle has minimal effect in this study. The optimal electrical value of 0.043 λ_{eff} (286 μm) is a suitable width, S_1, for this application because it gives the highest directivity and F/B ratio. This parameter may possibly have a smaller value at lower frequencies where the wavelength is much larger.

Table 13. Performance effects upon variation of width, S_1.

S_1 (μm)	Directivity (dBi)	Main beam tilt angle (deg)	F/B Ratio (dB)
286	11.4	35.5	15
886	11.3	35.3	11
1486	11.0	36.2	7
2086	10.2	37.4	5
2686	9.0	37.5	2

8.4.3 Variation of gap, g, at 32.7 GHz

The gap, g, is a very important parameter to analyze because it affects the coupling of the patch elements. Values of various gaps are summarized in Table 14. As this value increases, the directivity, scan angle, and F/B ratio is affected. The directivity decreases by as much as 2.3 dBi, while the F/B ratio severely diminishes by as much as 13 dB as the gap size is increased from 140 μm to 300 μm. The decrease in F/B ratio is due to the fact that the major contribution of radiation comes from the driven element and edge diffraction effects from a large-size finite substrate that cause the main beam to split into two lobes with a small F/B ratio (2-3 dB). Additionally, as g is increased, the main beam tilt angle tends to increase as well since the broadside radiation of the driven element becomes more intense and a second lobe is present in the radiation at broadside. The formation of this lobe gradually pushes the maximum radiation of the quasi-endfire main lobe to a higher angle. When g becomes too large (> 0.028 λ_{eff}), the presence of the D1 elements have no effect on the driven element, and therefore, only the driven element is radiating. The difference between 0.021 λ_{eff} (140 μm) and 0.045 λ_{eff} (300 μm) in

Table 14. Performance effects upon variation of gap, g.

g (μm)	Directivity (dBi)	Main beam tilt angle (deg)	F/B Ratio (dB)
140	11.4	35.5	15
190	11.1	37.2	10
240	10.5	39.9	6
300	9.2	44.2	2

selecting the gap width can be quite critical. As shown in Table 14, a 50 μm difference in the gap size (140-190 μm) results in a 5 dB difference in the F/B ratio and a difference in the gain of the antenna. To get a better idea of the F/B ratio and gain of the antenna at smaller intervals of change in the gap size as well as identifying the sensitivity of this parameter on the design, simulations were conducted starting at a gap size of 140 μm with a +5 μm increment up to 200 μm. In the analysis, it was shown that at 140 μm, a directivity of 11.4 dBi and F/B ratio of 15 dB is given. At 200 μm, a directivity of 11.0 dBi and F/B ratio of around 9 dB is observed. At all gap sizes in between 140 and 200 μm, the directivity and F/B ratio decreases monotonically from g = 140 μm to 200 μm. Over-etching can diminish the desired gain and desired F/B ratio. It is also worth noting that a 60 μm increase in the gap from the desired value of 140 μm still results in a 10 dB F/B ratio which is improved in comparison to past research of these antenna types.

8.4.4 Effects on Bandwidth Variation

The bandwidth of the design is mostly controlled by the S_1 parameter which is the spacing between the D1 elements. This is because the D1 elements are also resonant (as well as the driven element). When the coupling between the elements (driven and D1s) is strong, the D1 elements are excited electromagnetically. Since the resonant lengths of the driven and D1 elements are close in value, two TM_{10} resonance peaks are present, which are also close in frequency, and the bandwidth of the antenna is significantly larger (close to 8%). On the other hand, as the value of S_1 becomes larger, the coupling is reduced to a point where there is no coupling between the driven and D1 elements. Therefore, the driven element is the only resonant element present, and hence, the bandwidth is suppressed (by about 4%) due to the elimination of the TM_{10} mode of the

D1 element. This is seen in Figure 85 at three values of S_1. As S_1 increases, the higher resonance gradually decreases to a point where only the driven patch resonance is present. The effect of S_1 serves to steer the main beam as well as broaden the bandwidth of the design.

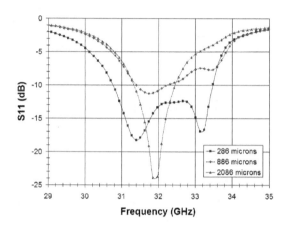

Figure 85. Return loss versus frequency at three values of S_1.

8.5 Simulation and Measurement Results

To verify the analysis that has been presented for this antenna structure, a scaled design is simulated around a band above and below 5.2 GHz for HIPERLAN applications. This design was fabricated by Prototron Circuits, Inc. and measured in the high frequency laboratory of Technology Square Research Building (TSRB) at Georgia Tech.

For the return loss measurement, an edge-mount SMA connector was soldered to the edge of the board, and the return loss was measured on a network analyzer. The simulated and measured return loss versus frequency is presented in Figure 86. (The physical dimensions of this design are shown in the caption of Figure 86.) There are two

frequency shifts in the resonances of the measured design when compared to the

simulated results. The lower resonance has a frequency shift of 20 MHz which is small

for this design, but the frequency shift of 80 MHz at the higher resonance can possibly be

attributed to a small tolerance in the substrate (RT/Duroid 5880, ε_r =2.2±0.02) and/or

over-etching of the metal. Additionally, the return loss at the resonances is higher (more

reflection), but definitely acceptable for this application. The connector was not taken

into account in the simulation. The bandwidth of the measured prototype (10%) is

considerably larger than the simulated design (7.8%). This can be mainly attributed to

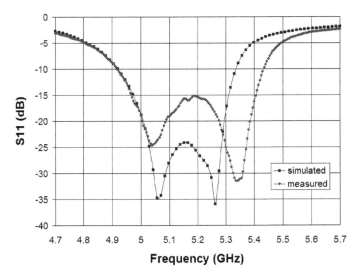

Figure 86. Return loss versus frequency for prototype at a band around 5.2 GHz. Physical dimensions are as follows: L_R = 245 mils, W_R = 1002 mils, L_D = W_D = 724 mils, L_{D1} = L_{D2} = 688 mils, W_{D1} = W_{D2} = 492 mils, S_1 = 72 mils, S_2 = 902 mils, g = 35 mils. Substrate: RT/Duroid 5880 (ε_r =2.2, tan δ =0.0009 @ 10 GHz) with thickness = 62 mils.

the shift in frequency at the higher resonance. It is worth noting that the broadened

bandwidth of the measured design may not be suitable for narrowband applications. The

size of the ground plane is $1.95\lambda_0$ x $1.95\lambda_0$.

To analyze the effect of the ground plane on the structure, the ground plane was

increased 12 times up to a factor of 120% in increments of 10% of the current size in both

directions (x and y). This study showed that as the ground plane increases from 10% to a

factor of 40% of the current size, the directivity decreases by small increments (about 0.3

dBi) and the F/B ratio decreases from around 15 dB to 11-12 dB. Ground size

enlargements from 50% to 120% result in the main beam becoming degraded to a point

where the beam is split into two. This is largely due the edge diffraction effects that

occur at the edges of the substrate when the ground plane is large.

The radiation pattern was measured in the anechoic chamber of TSRB. The

antenna was mounted on a metallic platform by using tape and connected to a coaxial

cable. The antenna-under-test (AUT) served as the transmitting antenna; while, a WR159

(4.9-7.05 GHz) rectangular horn antenna was used to receive the wave. The E-plane

radiation pattern, shown in Figure 87, was measured at 5.225 GHz and compared to the

simulated results. Due to the design of the chamber, only an E-plane measurement in the

range of $-90°\leq \theta \leq 90°$ could be taken. This is sufficient to compare the main beam and

the backside radiation level $(-90°\leq \theta \leq 0°)$. The co-polarized components of the measured

and simulated designs are in good agreement with each other. The main beam has been

shifted by 8°. This may be attributed to the stronger concentration of fields on the D1

elements as opposed to the driven element. The measured F/B ratio is 15 dB which is a

significant improvement over other designs presented in past literature. The measured

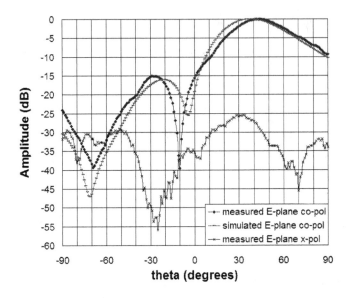

Figure 87. E-plane radiation patterns in rectangular form at 5.2 GHz.

beamwidth is 38° (26° ≤ θ ≤ 64°). This is due to the sharp decline in amplitude between 12-18°. The gain of the antenna is 10.7 dBi with an efficiency greater than 85%. The cross-polarized component of the measured design is 25 dB below the main beam.

8.6 Comparison with Two Yagi Arrays Separated by a Finite Distance

It is interesting to compare the results of placing two Yagi arrays next to each other separated by a finite distance (Figure 88) and the current Yagi design shown in Figure 78. The lengths of the elements, in Figure 88, are as follows: reflector length = 1002 mils, driven length = 724 mils, directors 1 and 2 = 688 mils, element spacing = 35 mils, and center-to-center driven element spacing = 1386 mils (0.61 λ_0 @ 5.2 GHz). This design is similar to one proposed in [79] that has four Yagi arrays placed next to each other separated by a finite distance. The electrical spacing between the arrays is the same

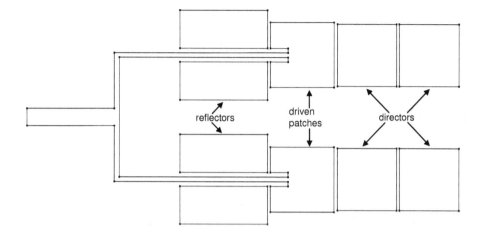

Figure 88. Antenna structure of two Yagi designs side-by-side separated by a finite distance.

as that in [79]. Through simulations, it is observed that a slightly higher directivity (11.9 dBi) can be obtained in the Figure 88 design (only 11.6 dBi for Figure 78 design). The E-plane radiation pattern is shown in Figure 89 for both designs. Despite a favorable directivity in the conventional two-Yagi design, the co-polarized component shows that the F/B ratio is much smaller than that obtained in the single driven element Yagi design (16 dB). The major reason why the F/B ratio is increased significantly is attributed to the gap between the two individual Yagi arrays that leads to the increased backside radiation. A few other variations of utilizing a continuous reflector plane and a continuous grounded reflector plane to minimize the effect of the gap was attempted. The directivity can be increased by as much as 0.3 dBi, but the improvement in the F/B ratio is negligible. This is the major advantage of this new design. One can obtain a high gain, quasi-endfire radiation pattern with a low cross-polarization and a high F/B ratio.

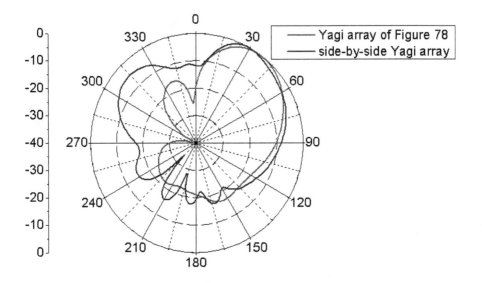

Figure 89. E-plane co-polarized component of side-by-side Yagi array (Figure 88) and current Yagi design (Figure 78) at 5.23 GHz.

8.7 Sensitivity Analysis and Optimization Using Design of Experiments and Monte Carlo Simulation

A sensitivity analysis and optimization was conducted for the antenna configuration in Figure 78 designed to operate at 5.2 GHz using DOE and Monte Carlo simulation to analyze the performance of the microstrip Yagi antenna array. In this section, only a summary of the results is discussed; the reader is referred to Section 5.6 for greater background and in depth detail about these methods.

The variables that affect this design are chosen to be the horizontal gap, g, between the elements, the distance between the D1 elements, S_1, and the distance between the D2 elements, S_2. The analysis intervals for the three variables are presented in Table 15.

Table 15. Ranges for the microstrip Yagi design input variables.

	g (mils)	S_1 (mils)	S_2 (mils)
"-" level	20	30	600
"+" level	60	130	1100
Center point	40	80	850

Since the statistical models are based on deterministic simulations, the variation of the center points were statistically simulated based on a ± 10 % tolerance and a 3σ fabrication process for all inputs, given by the tolerances of the RT/Duroid 5880 substrate material. The next step is the development of the transfer functions that utilize statistically quantified design parameters to predict the nominal values and process-based variations of the three figures of merit, which are the directivity (D), the F/B ratio, and the main beam tilt angle (TA).

The statistical analysis of the first order models shows which effects and interactions between the factors are significant for each of the three figures of merit, and those that are not significant are eliminated from the final models. In this analysis, curvature has not been detected and the first order models, which were investigated for the normality and equal variance assumptions, are presented below:

$$D=11.29-0.24\left(\frac{g\text{-}40}{20}\right),\qquad (35)$$

$$F/B=13.51-3.27\left(\frac{g\text{-}40}{20}\right)+0.25\left(\frac{S_2\text{-}850}{250}\right)-\left(\frac{g\text{-}40}{20}\right)\left[3.46\left(\frac{S_2\text{-}850}{250}\right)\right], \quad (36)$$

and

$$TA = 36.15 + 1.18\left(\frac{g\text{-}40}{20}\right).$$ (37)

Then, the structure was optimized based on the models for the following goals: maximum directivity, maximum F/B ratio and a tilt angle between 35-37°. The optimal values are found to be 11.34 dBi, 17.33 dB, and 35.9°, respectively, for the combination of inputs, g = 36 mils and S_2 = 1100 mils. These values coincide well with those obtained through measurements. The value of the F/B ratio is slightly larger than that obtained via measurements (15 dB), and the directivity is approximately 0.3 dBi smaller than that acquired in electromagnetic simulation. It is interesting to note in the equations that the S_1 parameter did not have a significant effect on the directivity, F/B ratio, or the tilt angle for the range considered in this section. The author performed an investigation in which the S_1 was varied to values greater than 130 mils. Three additional electromagnetic simulations were performed to look at the effect of S_1 when its value is varied to 150, 190, and 230 mils. The values for the gap, g, and width, S_2, are given in the caption of Figure 86. Table 16 summarizes the results of the three figures of merit. From this table, it is seen that the figures of merit exhibit little change based on S_1. Therefore, it is concluded that the statistical models are indeed consistent with the results based on the variation of S_1 in the range of 30-130 mils.

Table 16. Performance of figures of merit upon variation of width, S_1.

S_1 (mils)	Directivity (dBi)	F/B Ratio (dB)	Main beam tilt angle (deg)
150	11.58	13.0	36.9
190	11.57	12.5	36.1
230	11.56	11.9	35.8

The performance capability of the system was evaluated for the optimal structure

using Monte Carlo simulation. Table 17 shows the results of the Monte Carlo simulation.

Table 17. Predicted performance of the outputs of the microstrip Yagi design.

	D (dBi)	F/B (dB)	TA (deg)
Nominal	11.34	17.33	35.9
USL	n/a	n/a	36.51
LSL	11.21	15.45	35.80
Cp	n/a	n/a	2.03
Cpk	1.52	1.59	1.51

The first row shows the nominal values of the outputs obtained by plugging in the

optimized values of the inputs into the models. USL and LSL are the upper and lower

specification limits, respectively, and they represent the worst case scenario for each of

the outputs. Cp and Cpk are metrics that quantify evidence that the system complies with

six sigma process capability. Six sigma capability is reached for processes that achieve

$Cp \geq 2$ and $Cpk \geq 1.5$ for processes with USL and LSL, and $Cpk \geq 1.5$ for processes with

only USL or LSL, allowing, in both cases, the possibility of a long-term ±1.5 sigma shift.

In this case, Table 17 shows that these conditions are satisfied, and six sigma capability,

which includes the possibility of a long-term ±1.5 sigma shift, is reached.

8.8 Extension to Microstrip Bi-Yagi and Quad-Yagi Antenna Arrays

The proposed microstrip bi-Yagi and quad-Yagi antenna array designs are displayed in Figures 90 and 91, respectively. These designs are a derivative of the original microstrip Yagi antenna array in which a high gain is obtained through the constructive interference of two individual microstrip Yagi structures (R-D-D1$_T$-D2$_T$ and R-D-D1$_B$-D2$_B$) that maximally radiate at $10° \leq \varphi \leq 16°$ and $-16° \leq \varphi \leq -10°$, respectively. (An illustration of this antenna is shown in Figure 78.) The operational frequency is around 5.2 GHz, but frequency scaling of this design is quite simple since the only manufacturing tolerance is the minimum trace of the metals (no via processing required). The major benefit of these structures, compared to the structure in [85], is the increased gain (by 3-6 dBs) that can be achieved based on the design of the antenna. The common dimensions of the antennas in Figures 90 and 91 are as follows: the length and width of the reflectors, R, is ($L_R =$) 245 x ($W_R =$) 1002 mils, the length and width of the driven element, D, is ($L_D =$) 724 x ($W_D =$) 724 mils, and the lengths and widths of the director1 (D1) and director2 (D2) elements are ($L_{D1} = L_{D2} =$) 688 x ($W_{D1} = W_{D2} =$) 492 mils. The gap between the elements along the x-axis (g) is 35 mils. These values are chosen to optimize the F/B ratio and the gain of the antenna. Furthermore, the distances between the director1 and director2 elements are represented by S$_1$ and S$_2$, respectively. S$_1$ is 72 mils and S$_2$ is 902 mils. The small value of S$_1$ is due to the necessity of enhancing the coupling between the driven element and the D1 elements; while conversely, the larger value for S$_2$ is due to the necessity for a large aperture width to achieve an increased gain. These values were chosen to optimize the F/B ratio and the gain of the antenna. In the microstrip bi-Yagi array design, two conventional microstrip Yagi arrays have been

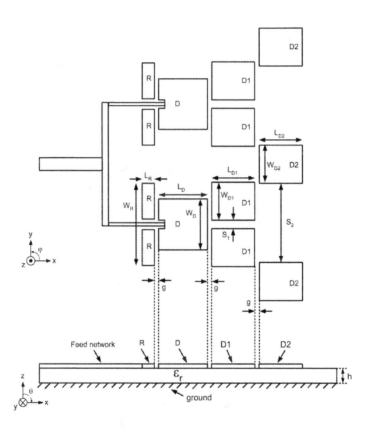

Figure 90. Illustration of microstrip bi-Yagi antenna array.

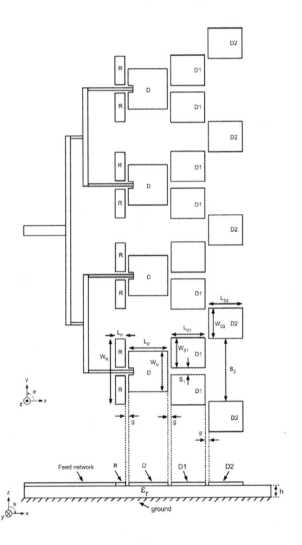

Figure 91. Illustration of microstrip quad-Yagi antenna array.

connected in a composite array format where a common D2 element has been used to prevent H-plane sidelobes that can arise when the center-to-center separation between the driven elements is greater than one wavelength (λ). Although the center-to-center spacing between the driven elements is 1548 mils (1.5 λ), the H-plane sidelobes are 17 dB below the main beam. Conversely, in the microstrip bi-Yagi array design, four conventional microstrip Yagi arrays have been connected in an array format where three common D2 elements are used to connect pairs of microstrip Yagi arrays. The center-to-center spacing between the driven elements is also 1548 mils. A smaller value for center-to-center spacing (1.0-1.5 λ) can further minimize sidelobes in the H-plane, but the gain becomes smaller. On the other hand, larger values (greater than 2.0 λ) can greatly increase the gain at the expense of obtaining a sidelobe level (SLL) less than 8 dB. So, this value (1.5 λ) is a tradeoff between achieving a minimal SLL and a high gain. The size of the substrate for the microstrip bi-Yagi array is 4560 x 5400 mils, while the quad-Yagi array has a size of 4940 x 8496 mils. The size of the substrate in the y-direction increases by 1548 mils each time a microstrip Yagi antenna array is added to the larger array to support the antenna. Both antenna structures are designed on a double copper clad board of RT/duroid 5880 material. The thickness of the substrate (h) is 62 mils. A thicker substrate could lead to a larger bandwidth, but the Yagi effect of quasi-endfire radiation would be degraded considerably due to the surface waves in the substrate. To maintain simplicity in the fabrication of the double copper clad board, gaps are inserted between the reflector patches, and the feedlines are connected to the driven patch through gaps in the reflectors.

8.9 Simulated and Measured Results of Bi-Yagi and Quad-Yagi Antenna Arrays

The microstrip bi-Yagi and quad-Yagi antenna arrays were simulated using MicroStripes 7.0. After an optimized design was obtained, the two antennas were fabricated by Prototron Circuits, as shown in Figure 92. Figure 93 displays the simulated return loss versus frequency of the bi-Yagi and quad-Yagi arrays compared to that of the original (one branch) microstrip Yagi array design. The bandwidths of the three designs are as follows: microstrip Yagi - 8.1%, bi-Yagi - 7.1%, and quad-Yagi - 5.0%. In comparing the three designs, it is apparent that the bandwidth tends to decrease as more microstrip Yagi arrays are added to produce the larger array. The smaller bandwidth of the quad-Yagi array may be due to the shift of the lower resonance of the driven element to a higher frequency. The measured return loss plots versus frequency of the antennas are also displayed in Figure 93. The measured bandwidths of the bi-Yagi and quad-Yagi designs are 6.9% and 5.2%, respectively. Although the lower and higher resonances of the measured designs occur at similar frequencies, the return loss of the quad-Yagi array is higher than that of the bi-Yagi array, resulting in a smaller bandwidth.

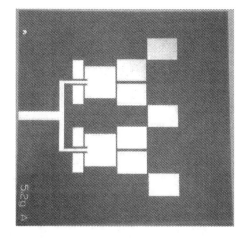

a.

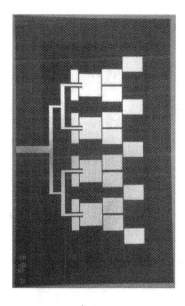

b.

Figure 92. Illustration of fabricated a.) microstrip bi-Yagi and b.) microstrip
quad-Yagi antenna array.

173

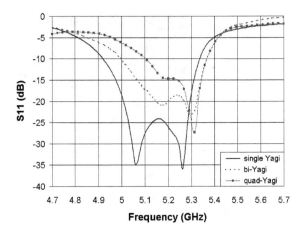

a.

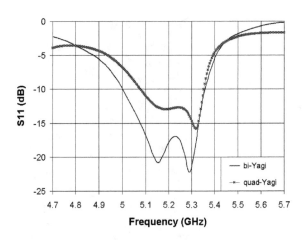

b.

Figure 93. a.) Simulated return loss of the three Yagi arrays and b.) measured return loss
of the designs proposed in this paper.

The simulated (normalized) 2D radiation patterns comparing the three Yagi designs at 5.2 GHz are presented in Figure 94. From this plot, it is observed that the F/B ratio tends to decrease as more Yagi arrays are included to produce the larger array. At some frequencies close to 5.2 GHz, the F/B ratio can be increased as the cost of a lower gain (by 0.5 dB); hence, there is a tradeoff. Tables 18 and 19 show how the gain and F/B ratio varies with frequency.

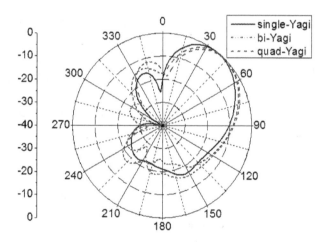

Figure 94. Simulated 2D radiation patterns of the three Yagi arrays at 5.2 GHz.

Table 18. Variation of Gain and F/B ratio versus frequency of bi-Yagi array.

Frequency (GHz)	Gain (dBi)	F/B Ratio (dB)
5.1	12.5	14
5.2	13.0	12.5
5.3	13.3	10.7

Table 19. Variation of Gain and F/B ratio versus frequency of quad-Yagi array.

Frequency (GHz)	Gain (dBi)	F/B Ratio (dB)
5.1	15.3	16
5.2	15.6	14
5.3	16.0	11.5

The angle of maximum radiation for all the designs is between 35-45°, while the beamwidth coverage is approximately 40°. Figures 95 and 96 display the measured (normalized) 2D radiation patterns of the microstrip bi-Yagi and quad-Yagi arrays at 5.2 GHz. A good agreement is observed between the simulated and measured results in the bi-Yagi array although the measured design has a slightly lower F/B ratio (10 dB) in comparison to simulation. The gain is 13.0 dBi and the cross-polarization is below -25 dB. The quad-Yagi array also exhibits a good agreement between the simulations and measurements. For this design, a gain of 15.6 dBi can be obtained with a cross-polarization below -18 dB. Considering that these structures use a highly conductive metal (Cu) that was printed on a low loss dielectric, the efficiencies of all the Yagi designs presented are greater than 89%.

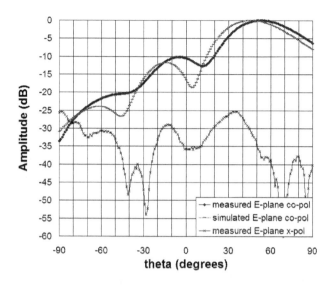

Figure 95. Measured 2D radiation pattern of the bi-Yagi array at 5.2 GHz.

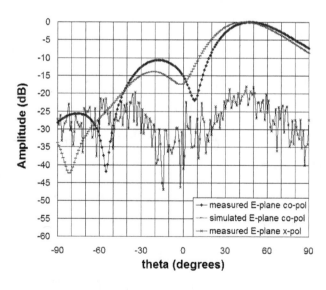

Figure 96. Measured 2D radiation pattern of the quad-Yagi array at 5.2 GHz.

CHAPTER 9

RF CERTIFICATES OF AUTHENTICITY

Counterfeiting is as old as the human desire to create objects of value.   For example, historians have identified counterfeit coins just as old as the corresponding originals. There are examples of counterfeit coins netting a 600% instant profit to the counterfeiter [86].   Test cuts were scheduled to be the first counterfeit detection procedure with an objective to test the purity of the inner structure of the coin.   The appearance of counterfeit coins with already engraved fake test cuts initiated the cat-and-mouse game of counterfeiters with original manufacturers that has lasted to date [86]. It is hard to assess and quantify the market for counterfeit objects of value today. With the ease of marketing products on-line, it seems that selling counterfeit objects has never been easier. National economies and industries that are under attack include the software and hardware, the pharmaceutical, the entertainment, and the fashion industry. According to a 2000 study by International Planning & Research, software piracy resulted in the loss of 110,000 jobs in the US, nearly US$1.6B in tax revenues and US$5.6B in wages. Similarly, pharmaceutical companies are commonly referring to the fact that over 10% of all medications sold worldwide are counterfeit. Consequently, there exists a demand for technologies that can either resolve these problems or significantly reduce the breadth of the search space for origins of piracy. Certificates of authenticity (COAs) aim at this objective.

A COA is a digitally signed physical object that has a random unique structure which satisfies three requirements: the cost of creating and signing original COAs is small, relative to a desired level of security, the cost of manufacturing a COA instance is several orders of magnitude lower than the cost of exact or near-exact replication of the unique and random physical structure of this instance, and the cost of verifying the authenticity of a signed COA is small, again relative to a desired level of security. An additional requirement, mainly impacted by a desired level of usability, is that a COA must be robust to ordinary wear and tear. COA instances can be created in numerous ways. For example, when covering a surface with an epoxy substrate, its particles form a lowrise but random 3D landscape which uniquely reflects light directed from a certain angle. COAs based upon this idea were first proposed by Bauder and Simmons from the Sandia National Labs and used for weapons control during The Cold War [87].

This chapter presents some initial analysis that has been done to define certificates of authenticity in the RF domain and quantify its importance and potential uses in consumer and defense applications.

9.1 Physical Phenomena of COAs in the RF Domain and System Setup

While there may be numerous phenomena in the electromagnetic domain that can be exploited for the design of RF COAs, the key new concept that is introduced is the effect of near-field measurements of electromagnetic properties exhibited by a COA instance. In general, electromagnetic fields are characterized by their electric field intensity, E, and magnetic field intensity, H. In material media, the response to the excitation produced by these fields is described by the electric flux density, D, and the magnetic flux density, B. The interaction between these variables is described using the Maxwell's equations:

$$\nabla \times H = \frac{1}{c} \frac{\partial D}{\partial t} + \frac{4\pi}{c} J$$

$$\nabla \times E = -\frac{1}{c} \frac{\partial B}{\partial t}$$

$$\nabla \cdot D = 4\pi\rho \qquad (35)$$

$$\nabla \cdot B = 0$$

where c is the speed of light in free space, and J and ρ denote the electric current density and charge density, respectively. (It is important to note that the equations in this chapter are based on the principles of theoretical physics. Hence, these equations are in "CGS" form and not in "MKS" form as is commonly used in the study of engineering.) For most media, we have linear relationships:

$$D = E + 4\pi P = \varepsilon E, \ B = \mu_0 H + M = \mu H, \ J = \sigma E \qquad (36)$$

where ε, μ, and σ are dielectric permittivity, magnetic susceptibility, and material's specific conductivity, respectively, and P and M are the polarization and magnetization vectors, respectively. From equations 35 and 36, one can derive additional equations that model the propagation of a monochromatic electromagnetic wave:

$$F_e = \nabla \times \nabla \times E - k^2 E$$

$$F_e = -4\pi \left(\frac{ik}{c} J + k^2 P + ik\nabla \times M \right) \qquad (37)$$

and

$$F_m = \nabla \times \nabla \times H - k^2 H$$

$$F_m = 4\pi \left(\frac{1}{c} \nabla \times J + k^2 M - ik\nabla \times P \right) \qquad (38)$$

where $k = \omega/c$ is the wavenumber. Equations 37 and 38 fully describe electromagnetic waves in 3D space. However, another form is commonly used for simulation of scattering based on the Ewald-Oseen extinction theorem [88-89]. Consider a material medium occupying a volume, V, limited by a surface S and use $r_>$ and $r_<$ to denote vectors to an arbitrary point outside and inside V respectively. An illustration is shown in Figure 97.

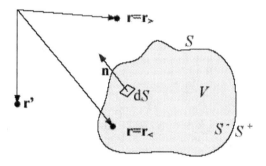

Figure 97. Material medium occupying volume, V, limited by surface, S.

The dyadic form, $G(r, r')$, of the scalar Green function, $G(r, r')$,

$$G(r,r') = \left(\mathcal{I} + \frac{1}{k^2} \nabla\nabla \right) G(r,r')$$

(39)

and

$$G(r,r') = \frac{\exp\left(ik|r-r'| \right)}{|r-r'|} ,$$

(40)

where $\mathcal{G}$ is a unit dyadic, describes a spherical wave at point r sourced from point r'. Now, the generalized extinction theorem [90] states:

$$E(r_<)=\frac{1}{4\pi}\int_V F_e(r')\cdot G(r_<,r')d^3r'-\frac{1}{4\pi}\sum_e^{(-)}(r_<), \tag{41}$$

$$E^{(i)}(r_<)+\frac{1}{4\pi}S_e(r_<) = 0, \tag{42}$$

$$E(r_>)=E^{(i)}(r_>)+\frac{1}{4\pi}S_e(r_>), \tag{43}$$

and

$$0=\frac{1}{4\pi}\int_V F_e(r')\cdot G(r_>,r')d^3r'-\frac{1}{4\pi}\sum_e^{(-)}(r_>), \tag{44}$$

where points r and r' are both inside of V (equation 41), inside and outside of V (equation 42), both outside of V (equation 43), and outside and inside of V (equation 44), respectively. $E^{(i)}$ is the incident field upon V and

$$S_e = \int_{S^-}\left[\left(n\times(\nabla\times E-4\pi ikM)+\frac{4\pi ik}{c}J\right)\cdot G(r,r')+\left(n\times E\right)\cdot\nabla\times G(r,r')\right]dS \tag{45}$$

$$\sum_e^{(-)}(r_<) = \int_{S^-}\left[\left(n\times\nabla\times E\right)\cdot G(r,r')+\left(n\times E\right)\cdot\nabla\times G(r,r')\right]dS \tag{46}$$

182

where S⁻ signifies integration approaching the surface S from the inside of V and n is a outward unit vector that is normal to dS. An analogous set of equations can be derived for the magnetic field [90]. Equations 42 and 43 and their magnetic analogues are particularly important because they govern the behavior of the electromagnetic field inside and outside of V when the source is outside of V. They can be restated in various forms which can be adjusted to alternate material conditions (non-magnetic, non-conductor, linear, isotropic, spatially dispersive, etc.). Providing numerical solutions to these equations is not a simple task in particular when the field values are computed in the near-field of the scatterers. Most research in the field targets radar, communication, and geodesic applications, which focus on approximating rough surfaces with a Gaussian distribution and computing the first and second order statistics of the exerted electromagnetic far-field [91-95]. For an arbitrary field setup, one likely must revert to one of the classical electromagnetic field equation solvers that address the above equations 42 and 43. There are numerous methodologies used for finding approximate solutions of partial differential equations as well as of integral equations: FDTD [45], MOM [46], and FEM [47]. Commercial simulators typically offer several solvers with distinct advantages for certain problem specifications [96]. In general, the computational complexity for most techniques is linked to their accuracy; accurate methodologies are typically superlinear: $O(N \log N)$ for improved MOM and FEM [97] and $O(N^{1.33})$ for FDTD [98], where N equals the number of discrete elements (typically, simple polygon surfaces) used to model the simulated electromagnetic environment. In the system that is proposed, for a known RF COA topology, it is anticipated that accurate simulations would require in excess of N > 108 discrete elements. An example of the substantial discrepancy in accuracy and performance of modern field solvers can be observed in a recent comparison study of six state-of-the-art solvers [99]. For a relatively simple semi-2D vivaldi antenna with an operating frequency at 4.5 GHz, modeled with approximately $N \sim 105$ discrete elements, the S_{21} response of a single simulation in the 3-7 GHz band differed by up to 12 dB with substantial differences with respect to the actual measurements of the physical implementation of the structure. The fastest program in the suite returned accurate results after approximately one hour on an 800 MHz Pentium processor. In summary, after research was conducted in this important field, it is the author's opinion that state-of-the-art tools are far from fast and accurate.

The key to system efficiency is to produce a reader capable of reliably extracting an RF "fingerprint" from a COA instance in the frequency range between 5-6 GHz. In order to disturb the near-field of the COA instance, the instance is constructed as a collection of randomly bent, thin conductive wires with lengths randomly set within 3-7 cm. The wires are integrated into a single object using a transparent dielectric sealant, illustrated in Figure 98. The sealant fixes the wires' positions within a single object. The "fingerprint" of such a COA instance should represent the 3D structure of the object. To address this issue, a reader designed as a matrix of antennas is proposed with an analog/digital back-end. Each antenna can behave as a transmitter or receiver of RF waves in a specific frequency band supported by the back-end processing. For different constellations of dielectric or conductive objects between a particular transmitter/receiver coupling, the scattering parameters for this coupling are expected to be distinct. Hence, to compute the RF "fingerprint," the reader collects the scattering parameters for each transmitter-receiver coupling in the group of antennas. The measurements of the reader

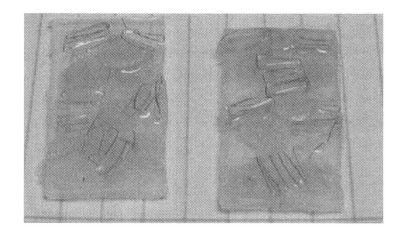

Figure 98. Copper wires and dielectric sealed in a transparent rubber mold.

represent the electromagnetic effects that occur in the near-field of the transmitter, the COA instance, and the receiver; in addition, the distances between any two objects are proportional to the wavelengths of interest. The near-field electromagnetic effects are observed for several reasons: it is hard to maliciously jam near-field communications, the reader can operate with low-power, low-efficiency antenna designs, and the variance of the electromagnetic field is relatively high in the near-field, causing better distinguishing characteristics. Far-field responses typically represent certain average characteristics of random discrete scatterers [92]; thus, such responses lose the ability to represent the scatterer's random structure.

A prototype RF COA scanner is designed as a matrix of 5x10 antennas that measures the unique RF "fingerprint" of an RF COA instance as a collection of transmission responses in the 5-6 GHz frequency range for each transmitter-receiver coupling on the reader. RF COA instances were placed at about 0.5 mm from the antenna matrix, i.e., in the near-field of the scanner. While the analog/digital back-end in the testbed was resolved using an off-the-shelf network analyzer, it is speculated that a custom reader could cost less than $100 if manufactured en masse.

9.2 Issuance and Verification of COAs and Possible System Attacks

The process of issuing and verifying a COA is displayed in Figure 99. When creating an RF COA instance, the issuer digitally signs the instance's RF response using traditional cryptography as follows. First, the unique RF "fingerprint" is digitized and compressed into a fixed-length bit string f. Next, f is concatenated to the information t (f ‖ t) associated with the tag (e.g., product ID, expiration date, assigned value) to form a combined bit string $w = f ‖ t$. Message w is then hashed using a cryptographically-strong algorithm H(), such as SHA256 [100]; next, the hash is signed using the private key of the issuer to create a message $m = w ‖ S(H(w))$. Function S() is the signing primitive of an adopted public-key cryptosystem (PKCS), such as RSA [101]. Message m is encoded directly onto the COA instance using existing technologies such as 2D barcodes or RFIDs. In order to reduce the length of m, one possibility is to use a PKCS based upon elliptic curves [102]. Message m is used to validate in-field that the produced instance is authentic. Each RF COA instance is associated with an object whose authenticity the issuer wants to vouch. Once issued, an RF COA instance can be verified off-line and in-field by anyone using a trusted reader (a reader that contains the corresponding public key of the issuer). Verifying a COA instance involves the following steps. First, the verifier reads $m = w ‖ S(H(w))$ from the attached physical storage and verifies the integrity of w with respect to $S(H(w))$ using the issuer's public key and a verification

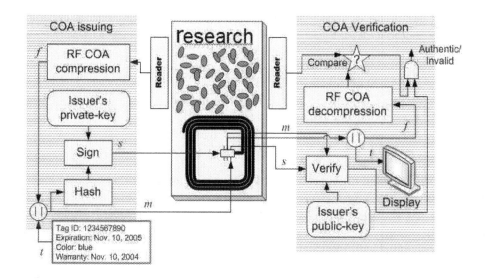

Figure 99. The process of issuing and verifying a COA.

primitive V() that corresponds to S(). In case the integrity test, V(w; S(H(w))), is successful, the original RF "fingerprint" f and associated data g are extracted from w.

The verifier proceeds to scan the actual RF "fingerprint" of the attached RF COA, i.e., obtain a new reading of the instance's RF properties, and compare them with f. If the level of similarity between f and f' exceeds a predefined and statistically validated threshold, the verifier declares the COA instance to be authentic and displays t. In all other cases, the reader concludes that the instance is not authentic, i.e., it is either counterfeit or erroneously scanned. To counterfeit protected objects, the adversary needs to compute the private key of the issuer (a task which can be made arbitrarily difficult by adjusting the key length of the used public-key crypto-system [100]), devise a manufacturing process that can exactly replicate an already signed COA instance (a task which is not infeasible but requires certain expense by the malicious party; the forging cost dictates the value that a single COA instance can protect [103]) or misappropriate signed COA instances (a responsibility of the organization that issues COA instances). From that perspective, COAs can be used to protect objects whose value roughly does not exceed the cost of forging a single COA instance including the accumulated development of a successful adversarial manufacturing process.

9.3 Design of Antenna and Prototype Scanner

To scan the electromagnetic features of an RF COA, the author proposes a scanner designed to expose the subtle variances of near field responses of these objects to RF waves. This RF COA scanner consists of a matrix of antennas, where each antenna is capable of operating both as a transmitter and a receiver. The antennas are multiplexed to an analog/digital back-end capable of extracting the S_{21} parameter (the insertion loss) for a particular antenna coupling. During the read-out, the antenna matrix is placed against the RF COA instance. The instance should have an absorbent and/or reflective background so that the environment behind the tag does not affect its RF response. By placing the instance in the close proximity of the antenna matrix as illustrated in Figure 100, one can collect numerous measurements including all S-parameters. For example, in a system with M antennas, one can measure M S_{11} and $\binom{M}{2}$ S_{21} parameters. The positioning of the instance with respect to the reader should be approximately the same for each reading. Depending on the accuracy of the analog and digital circuitry as well as the noise due to external factors, one can aim to maximize the entropy of this response.

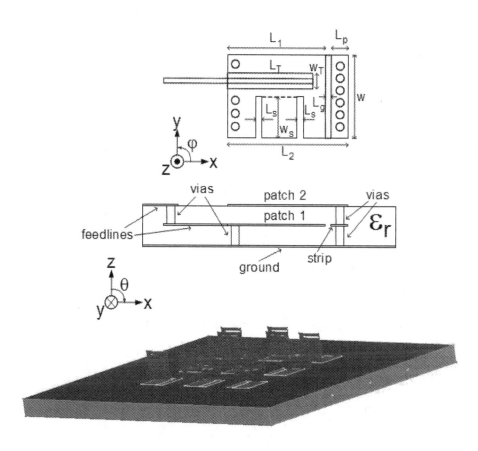

Figure 100. Antenna structure for antenna used in scanner and a 3x3 panel of antennas with instances above the surface.

A prototype scanner is designed in two steps. First, a microstrip antenna patch with an operating frequency close to the 5 GHz range is developed with an emphasis on miniaturization. In order to pack as many antennas as possible in a credit card size space, two minimization techniques were utilized: folding [41] and meandering [18]. Although the two techniques have already been applied to individual designs, to the best of the author's knowledge, this is the first study, both in simulation and implementation, which combines the two methods into a single design. The antenna is illustrated in Figure 100; its technical characteristics and design strategy are detailed in [104]. The major criterion in the return loss plot is the resonance of the antenna at a frequency around 5 GHz. In simulation of this system, the return loss is -16 dB at a resonant frequency of 4.93 GHz. The physical size of a single patch antenna that operates around the same frequency for a similar size substrate (RF60) is approximately five times larger than the design considered in this chapter. Based upon the simulated design, an extended panel prototype was constructed that consists of 50 antennas (five rows, ten columns). Prototron Circuits, Inc. fabricated the panels on RF60 substrate ($\varepsilon_r = 6.15 \pm 0.15$, $\tan\delta = 0.0019$) that had a total thickness of 62 mils. Fifty edge mount RF coaxial connectors were soldered to the ends of all the feedlines of the antennas. Transmission measurements of the antennas were performed using an Agilent 8753E vector network analyzer. The measurements were calibrated to the end of the coaxial cables. The S_{21} parameter was taken for many antenna couplings. The tester would manually attach and detach the ends of the cables to the connectors to switch to the different parameters. Figure 101 displays the prototype. The molds of Figure 98 were used as RF COA test-instances.

Figure 101. Fabricated 5x10 panel of antennas along with the test setup.

9.4 Empirical Evaluation

In the first set of experiments, an analysis aimed at quantifying the response's sensitivity to slight misalignment of the COA instance with respect to the reader antenna matrix is performed. In the second set, the goal is to estimate the response entropy as perceived by the verifier. Figure 102 illustrates the set of antenna couplings that are active during each experiment. Since a multiplexer for antenna ports was not constructed, the data collection process was limited to only a subset of all possible active couplings as each reading was manually calibrated. For each antenna coupling, data was collected for the following cases: an RF COA instance that consisted of Dragon Skin^{TM} only (denoted by I_1) and eight RF COA instances with copper wire embedded in Dragon Skin^{TM}

(denoted by I_2). Each RF COA instance is separated from the antenna matrix using a 0.5 mm thick slice of styrofoam. Tests with COA instances built using a mix of two dielectrics required a three times greater COA thickness to produce the same variance of the RF "fingerprint". Thus, the selected benchmark is the focus of this work.

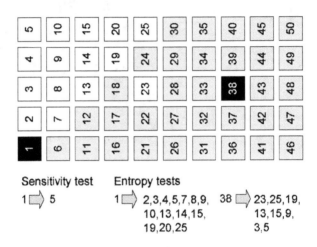

Figure 102. Schematic of fabricated 5x10 panel of antennas showing which antennas were used as a transmitter/receiver during alignment and entropy tests.

9.4.1 Alignment and Sensitivity Test

In the first experiment, antennas #1 (transmitter) and #5 (receiver) are activated and one of the I_2 instances is placed over the antenna matrix. Then, the responses are initially recorded and nine additional readings are taken each time by removing the instance completely from the matrix board and then manually realigning it to approximately the same position as in the initial reading. Positioning precision is on the order of 1 mm. The actual values and standard deviation of the resulting readings for the magnitude m_{21} and phase p_{21} of the complex response S_{21}, are illustrated in Figure 103. At lower frequencies and higher transmission efficiencies, the alignment variances $\sigma_m(f)^2$ = Var[$m(f)_{21}$] and $\sigma_p(f)^2$ = Var[$p(f)_{21}$] were substantially lower. Noticeable peaks in σ_m and σ_p were recorded towards the lower end of signal gaps at $f = \{5.5, 5.65, [5.85, 5.95]\}$ GHz. Within the range $f \in [5.85, 5.95]$ GHz, σ_m reached as high as 4.5 dB and the recorded response values are about 30 dB lower than response's peak P. It is important to note that weak response values should be proportionally ignored. The fact that σ_m is below 1.5 dB (mostly lower than 0.5 dB) for response values as low as P–20 dB is optimistic.

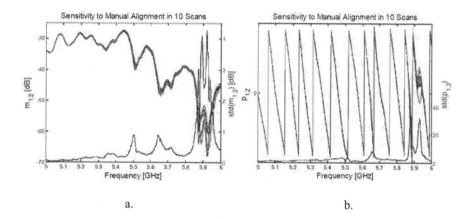

Figure 103. Plot of the a.) magnitude and b.) phase response of s_{21} along with their standard deviation.

Slight misalignment did not affect the phase information as critically as it affected the magnitude information. Finally, in practical situations where alignment is achieved mechanically with better than 0.1 mm precision, a significantly lower alignment variance is expected (σ_m<0.2 dB and σ_p<3). The only other significant source of noise in the detector can be attributed to manufacturing inaccuracies which may contribute an additional 0.3 dB of noise.

9.4.2 Entropy Test

Next, a set of experiments is conducted to evaluate the entropy of the conceived I_2 instance as identified by the developed scanner. Antennas #1 and #38 are activated as transmitters and a series of 14 and 8 antennas are used as receivers, respectively (as shown in Figure 102). Thus, a total of 22 couplings are considered which is a small subset compared to the total of 1225 possible couplings in the system. For each coupling, the response of a single I_1 instance and eight I_2 instances is recorded within the $f \in$ [5, 6] GHz frequency range at 1601 equidistant frequencies. To report the resulting entropy, the following steps are performed. For each of the eight I_2 instances ($i \in$ [1...8]), the differential response,

$$\hat{m}\{i\}(x,y) = m_{21}\{i\}(x,y) - m_{21}\{I_1\}(x,y), \tag{47}$$

is computed, where x and y denote the transmitter and receiver antenna respectively and {} encases the type of object scanned. The results for all the measured $\hat{m}\{i\}(x,y)$ responses are shown in Figure 104.

194

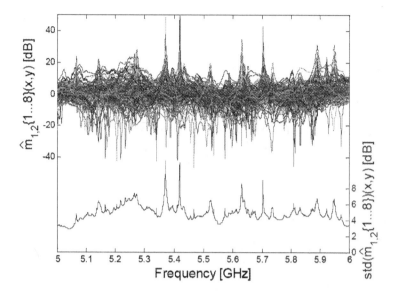

Figure 104. Plot of the magnitude differential response of s_{21}.

Next, the Euclidean distance,

$$d\{i,j\}(x,y)=\left\|\hat{m}\{i\}(x,y)-\hat{m}\{j\}(x,y)\right\|$$ (48)

is obtained for each $i,j \in [1...8]$ where $i \neq j$. Then, the distribution of $d\{i,j\}(x,y)$ is estimated for each coupling (x,y) using a χ^2-model and a maximum likelihood estimator. The estimated probability distribution curves,

$$\int_a^b \gamma_{x,y}(t)dt = Pr\left[E\left[d\{i,j\}(x,y)\right] \in [a,b]\right],$$ (49)

are computed for each individual antenna coupling (x,y). If one assumes that the noise margin for the readers when assessing the RF features of a COA instance is 0.5 dB, then for $d_T = 32$, a false negative error rate of $\varepsilon_{FN} \ll 10^{-6}$ is observed. Thus, the detection threshold is set at d_T and ε_{FN} is established. Then, the entropy of a single COA instance is estimated as perceived by the developed COA scanner as

$$H = -\log_2 \prod \forall (x,y) \in A \gamma x,y(d_T),$$
(50)

where A is the set of all antenna couplings. In this case, $|A| = 1225$. For antenna couplings (a,b) that were not measured, $\gamma_{x,y}$ is assumed such that the Euclidean distance between x and y is closest to the distance between a and b for the set A' of all measured responses (i.e., $(a,b) \in [A-A']$ and $(x,y) \in A$). Using this assumption, H is estimated to be 53832 bits. It is important to stress that while this entropy quantifies the likelihood of a false positive as $\log_2(\varepsilon_{FP}) = H$, it does not specify the difficulty of computing and manufacturing a false positive.

An RF "fingerprint" of an RF COA instance consists of a set of complex S_{21} parameters observed over a specific frequency band and collected for (a subset of) all possible antenna couplings on the reader. Each analog S_{21} parameter is sampled at arbitrary frequencies and individually quantized using an arbitrary quantizer. Signal f (introduced in Section 9.2) may consist of the raw data or a compressed RF "fingerprint". This compression may be lossy or lossless with respect to the digitized fingerprint extracted from a single instance.

9.5 Synopsis of Attacks

There are two potential key problems that need to be addressed in order to realize this technology. These problems are denoted as blind analysis and known-X manufacturing. The blind analysis is expressed as follows: given an RF "fingerprint", f, of an authentic RF COA instance extracted using a known RF COA scanner, find a three-dimensional object X capable of producing an RF response f' such that $\|f-f'\| < d_T$, where the detection threshold d_T is a proportionally small scalar. An additional requirement is to develop a manufacturing process that can produce X in large quantities at a relatively low price P. This is called the known-X manufacturing problem. As detailed in [103], in order for the counterfeiter to make profit, P must be smaller than the profit that the counterfeit product can fetch on the market. There are two layers of difficulty imposed upon the counterfeiter: a computational one (blind analysis) and a manufacturing one (known-X manufacturing). The author conjectures that both of these problems are difficult to answer and pose them as a security challenge to the scientific community. RF COAs can be used in scenarios where either one or both challenges are used to protect a physical object.

CHAPTER 10

CONCLUSIONS

The design, modeling, and optimization of compact broadband and multiband antenna architectures have been thoroughly investigated for use in many wireless applications. A thorough analysis of the theoretical performance, the schematic of the antenna structure, and the simulated and measurement results has been performed. A set of design rules has been established for the purpose of designing optimized bandwidth compact stack patch antennas on LTCC multilayer substrates. Some of the advantages of using LTCC are a lower dielectric loss, size reduction due to high dielectric constant, and the ability to incorporate embedded passives and interconnect circuitry to be sandwiched between the substrate layers. To verify its effectiveness, the proposed design rules have been applied to three emerging wireless bands: the 2.4 GHz ISM band, the IEEE 802.11a 5.8 GHz band, and the 28 GHz LMDS band. It has been observed that the return loss and the impedance bandwidth are optimized for all three bands. A maximum bandwidth of 7% in contrast to single patch's 4% has been achieved for an antenna operating in the LMDS band. The radiation patterns exhibit a similar performance in comparison to a single patch antenna, though, the cross-polarization levels have been significantly reduced, something that could enable the use of this antenna in 3G or millimeter-wave polarization-diversity systems. The derived design rules provide a detailed guide to constructing multilayer stacked patch antennas on LTCC or other organic (e.g. LCP, BCB) multilayer laminates that can be easily integrated with vertically integrated modules for a variety of frequency bands.

Folded shorted patch antennas (SPAs) can be easily implemented to significantly reduce the resonant frequency of a standard patch antenna. The design methodology of this structure starts as a conventional half-wavelength ($\sim\lambda_0/2$) antenna. From there, placing a metal wall along the middle line of the patch with a metal shorting wall reduces the resonant length to $\sim\lambda_0/4$. Then, a folding procedure that includes folding the ground and the patch simultaneously further reduces the resonant length to $\sim\lambda_0/8$. Upon completing this step, varying the height of the lower patch can result in as much as a $\sim\lambda_0/16$ resonant length. A comparison between a folded SPA and a standard SPA has validated the folding technique proposed in this book. Additionally, a theoretical analysis has been presented to justify the design methodology. The folded SPA has been applied to the 2.4 GHz ISM band and can be easily extended to higher frequencies using ceramic or organic substrates. The measured results have closely verified those obtained via simulation. This antenna can be implemented into 3D packages for wireless, automotive or miniaturized sensor applications.

To reduce the backside radiation due to surface wave propagation from a high dielectric substrate, the use of an SHS has been employed to the design of microstrip patch antennas. The SHS is realized by surrounding the patch with metal via rings whose height must be equal to $\lambda_0/(4*\varepsilon_r^{0.5})$. A reduced backside radiation of approximately 10 dB is achieved with the SHS in comparison to a similar design without the SHS at 64.55 GHz. This design can also achieve a gain enhancement of 10 dB at broadside (z-direction). The SHS has a broad coverage and a low cross-polarization level [1]. An improved design of an antenna surrounded by one soft surface ring has shown the same potential in suppressing surface waves and edge diffraction effects and in reducing backside radiation. With this implementation, a 5 dB improvement in the gain and an 8 dB improvement in the F/B ratio can be observed in comparison to a simple antenna design without the soft surface ring. Generally, a large ground plane can also be used to reduce backside radiation, but incorporating the soft surface ring to surround a patch antenna can significantly reduce the size of the antenna, therefore making for a more compact module.

A simple method of inverting negative resistances has been applied to the equivalent circuit modeling of dominant TM_{10} mode microstrip antennas. After a specific topology of the studied design is simulated using computational electromagnetic solving techniques, the admittance data is approximated by a rational function through a process called vector fitting. The passive elements are generated from the rational function and first applied to a patch antenna operating at 2.4 GHz as a benchmark design. When resistance invertibility is enforced, the equivalent circuit of this structure agrees well with the approximated function. To validate the methodology, this technique was further applied to the following resonant structures: a circular loop antenna and a dual resonant patch antenna with tuning capacitors. A good agreement between the equivalent circuits and the approximated function is observed. Future work will focus on applying these methods to generate circuits to fit the parasitic effects of complex antennas and resonant structures with accurate models, including the microstrip Yagi antenna array presented in Chapter 8. These techniques have the potential to be used in the understanding of electromagnetic structures by analyzing the electrical performance of the equivalent circuit for the frequency range of interest.

The development of two dual-frequency (14 GHz and 35 GHz), dual-polarization microstrip antenna arrays is presented for the first time using LCP multilayer technology. Some of the properties of LCP, such as multilayer (3D) vertical integration capability, good electrical and mechanical properties, and a near-hermetic nature, make this substrate a practical choice for the design of low cost, compact antenna arrays that can be integrated with remote sensing applications operating in the Ku and millimeter-wave frequency bands. After the simulations for the single frequency 2x1 and 2x2 designs were optimized, the arrays at each frequency have been integrated into a complete dual-frequency structure. RF MEMS switches can be used to switch polarizations, hence, introducing the possibility of realization of low-power reconfigurable antenna arrays. Full-wave simulations of the antenna designs have been performed and agree well with measured results of the scattering parameters and the radiation patterns.

A microstrip Yagi antenna array has been designed for various frequency bands to be integrated to ISM, IEEE802.11a (IEEE802.11n), and millimeter-wave applications. Bandwidths greater than 7.5% can be achieved through the use of the two close resonances of the driven and director elements. Quasi-endfire radiation in the range of 40° can be achieved by the configuration of this design. A high F/B ratio (15 dB) and a low cross-polarization are achieved that is suitable for applications where the backside level needs to be suppressed. Finally, a high gain (> 10dB) is observed through the use of multiple directors (four total: two D1 and two D2 elements) and the idea of constructive interference. To the author's knowledge, this is the first microstrip Yagi array design that has been proposed that can simultaneously achieve high gain quasi-endfire radiation with a high F/B ratio. In addition, two new antenna array designs based on the original microstrip Yagi antenna array have been discussed: the microstrip bi-Yagi array and the microstrip quad-Yagi array. Furthermore, the measured radiation pattern performance agrees well with the simulated results in terms of the beamwidth and maximum angle of radiation for both structures and gains of 13.0 and 15.6 dBi can be achieved with these designs with high efficiencies (above 89%).

Finally, the first system for manufacturing and verifying RF COAs which exhibit random behavior in the near-field has been investigated. A peculiar feature of this system, not exhibited in previous literature, is the conjectured difficulty of creating a COA instance that produces a specific response. A working prototype of the system has been demonstrated that has assisted in estimating system performance from the perspective of response repetitiveness and entropy. Based on the results, it is concluded that this technology has the potential of addressing many security problems that exists in applications, such as counterfeit tracing, consumer markets, and defense.

In summary, this book has described in depth the design process and analysis of antenna architectures that can be implemented into a wide range of applications, such as multilayer packages for the development of integrated wireless 3D SOP modules. Although these antennas are compact in size, the full functionality and performance capability of these designs have not been compromised. In fact, these designs show some performance improvement, such as a higher gain, lower cross-polarization, lower F/B ratio, and a smaller size. In addition, it is worth noting that compact antennas necessitate research in many interdisciplinary fields. The author has purposely presented this study in a way that solicits the input of the scientific community at large.

REFERENCES

1. R. R. Tummala, V. K. Madisetti, "System on chip or system on package," IEEE Design & Test of Computers, Volume: 16, Issue: 2, April-June 1999, pp. 48-56.

2. S. K. Lim, "Physical design for 3D system on package," IEEE Design & Test of Computers, Volume: 22, Issue: 6, November-December 2005, pp. 532-539.

3. R. R. Tummala, M. Swaminathan, M. M. Tentzeris, J. Laskar, G. –K. Chang, S. Sitaraman, D. Keezer, D. Guidotti, Z. Huang, K. Lim, L. Wan, S. K. Bhattacharya, V. Sundaram, F. Liu, P. M. Raj, "The SOP for miniaturized, mixed-signal computing, communication, and consumer systems of the next decade," IEEE Transactions on Advanced Packaging, Volume: 27, Issue: 2, May 2004, pp. 250-267.

4. R. R. Tummala, "SOP: what is it and why? A new microsystem-integration technology paradigm-Moore's law for system integration of miniaturized convergent systems of the next decade," IEEE Transactions on Advanced Packaging, Volume: 27, Issue: 2, May 2004, pp. 241-249.

5. R. R. Tummala, J. Laskar, "Gigabit wireless: system on package technology," Proceedings of the IEEE, Volume: 92, Issue: 2, February 2004, pp. 376-387.

6. K. Lim, S. Pinel, M. Davis, A Sutono, C. Lee, D. Heo, A. Obatoynbo, J. Laskar, M. Tentzeris, R. Tummala, "RF-system-on-package (SOP) for wireless communications," IEEE Microwave Magazine, Volume: 3, Issue: 1, March 2002, pp. 88-99.

7. R. Tummala, Fundamentals of Microsystems Packaging, New York, The McGraw-Hill Companies, Inc., 2001.

8. C. Balanis, Antenna Theory: Analysis and Design, New York, John Wiley & Sons, Inc., 1997.

9. W. Stutzman, G. Thiele, Antenna Theory and Design, New York, John Wiley & Sons, Inc., 1998.

10. D. Pozar, D. Schubert, Microstrip Antennas: The Analysis and Design of Microstrip Antennas and Arrays, New Jersey, IEEE Press, 1995.

11. R. Garg, P. Bhartia, I. Bahl, A. Ittipiboon, Microstrip Antenna Design Handbook, Boston, Artech House, Inc., 2002.

12. S. Consolazio, K. Nguyen, D. Biscan, K. Vu, A. Ferek, A. Ramos, "Low temperature cofired ceramic (LTCC) for wireless applications," IEEE MTT-S Symposium on Technologies for Wireless Applications, Feb. 1999, pp. 201-205.

13. L. Devlin, G. Pearson, B. Hunt, "Low-cost RF and microwave components in LTCC," Proceedings of MicroTech 2001, Jan. 2001, pp. 59-64.

14. D. C. Thompson, O. Tantot, H. Jallageas, G. E. Ponchak, M. M. Tentzeris, J. Papapolymerou, "Characterization of liquid crystal polymer (LCP) material and transmission lines on LCP substrates from 30 to 110 GHz," IEEE Transactions on Microwave Theory and Techniques, Volume: 52, Issue: 4, April 2004, pp. 1343-1352.

15. K. Wong, Compact and Broadband Microstrip Antennas, New York, John Wiley & Sons, Inc., 2002.

16. I. Bahl, P. Bhartia, Microstrip Antennas, Massachusetts, Artech House, Inc., 1980.

17. S. Bokhari, J. Zurcher, J. Mosig, F. Gardiol, "A small microstrip patch antenna with a convenient tuning option," IEEE Transactions on Antennas and Propagation, Volume: 44, Issue: 11, Nov. 1996, pp. 1521-1528.

18. S. Dey, R. Mittra, T. Kobayashi, M. Itoh, S. Maeda, "Circular polarized meander patch antenna array," IEEE Antennas and Propagation Society International Symposium, Volume: 2, July 1996, pp. 1100-1103.

19. K. Wong, J. Kuo, T. Chiou, "Compact microstrip antennas with slots loaded in the ground plane," IEEE Antennas and Propagation Society International Symposium, Volume: 2, July 2001, pp. 732-735.

20. H. Wang, M. Lancaster, "Aperture-coupled thin-film superconducting meander antennas," IEEE Transactions on Antennas and Propagation, Volume: 47, Issue: 5, May 1999, pp. 829-836.

21. M. Vaughan, K. Hur, R. Compton, "Improvement of microstrip patch antenna radiation patterns," IEEE Transactions on Antennas and Propagation, Volume: 42, Issue: 6, June 1994, pp. 882-885.

22. R. Mittra, S. Dey, "Challenges in PCS design," IEEE International Symposium 1999, Antennas and Propagation Society, Volume: 1, July 1999, pp. 544-547.

23. http://www.fcc.gov/cgb/sar/. April 2003.

24. R. Waterhouse, S. Targonski, D. Kokotoff, "Design and performance of small printed antennas," IEEE Transactions on Antennas and Propagation, Volume: 46, Issue: 11, Nov. 1998, pp. 1629-1633.

25. S. Targonski, R. B. Waterhouse, "An aperture coupled stacked patch antenna with 50% bandwidth," IEEE Antennas and Propagation Society International Symposium, Volume: 1, July 1996, pp. 18-21.

26. P. Bhartia, I. Bahl, "A frequency agile microstrip antenna," IEEE Antennas and Propagation Society International Symposium, Volume: 20, May 1982, pp. 304-307.

27. M. du Plessis, J. Cloete, "Tuning stubs for microstrip-patch antennas," IEEE Antennas and Propagation Magazine, Volume: 36, Issue: 6, Dec. 1994.

28. C. Huang, J. Wu, K. Wong, "Cross-slot-coupled microstrip antenna and dielectric resonator antenna for circular polarization," IEEE Transactions on Antennas and Propagation, Volume: 47, Issue: 4, April 1999, pp. 605-609.

29. H. Iwasaki, "A circularly polarized small-size microstrip antenna with a cross slot," IEEE Transactions on Antennas and Propagation, Volume: 44, Issue: 10, Oct. 1996, pp. 1399-1401.

30. P. Sharma, K. Gupta, "Optimized design of single feed circular polarized microstrip patch antennas," Antennas and Propagation Society International Symposium, Volume: 19, June 1981, pp. 19-22

31. K. Carver, J. Mink, "Microstrip antenna technology," IEEE Transactions on Antennas and Propagation, Volume: 29, Issue: 1, Jan. 1981, pp. 2-24.

32. W. Richards, "Microstrip antennas," Chapter 10 in <u>Antenna Handbook: Theory, Applications and Design</u>, New York, Van Nostrand Reinhold Co., 1988.

33. Y. Hwang, Y. Zhang, G. Zheng, T. Lo, "Planar inverted F antenna loaded with high permittivity material," Electronic Letters, Volume: 31, Issue: 20, Sept. 1995, pp. 1710-1712.

34. C. Huang, J. Wu, K. Wong, "High-gain compact circularly polarised microstrip antenna," Electronic Letters, Volume: 34, Issue: 8, April 1998, pp. 712-713.

35. G. Kumar, K. Gupta, "Broad-band microstrip antennas using additional resonators gap-coupled to the radiating edges," IEEE Transactions on Antennas and Propagation, Volume: 32, Issue: 12, Dec. 1984, pp. 1375-1379.

36. W. Hsu, K. Wong, "Broadband aperture-coupled shorted patch antenna," Microwave and Optical Technology Letters, Volume: 28, March 2001, pp. 1506-1508.

37. K. Wong, Y. Lin, "Small broadband rectangular microstrip antenna with chip-resistor loading," Electronic Letters, Volume: 33, Issue: 19, Sept. 1997, pp. 1593-1594.

38. E. Levine, G. Malamud, S. Shtrikman, D. Treves, "A study of microstrip array antennas with the feed network," IEEE Antennas and Propagation, Volume: 37, no.4, April 1989, pp. 426-434.

39. J. James, P. Hall, <u>Handbook of Microstrip Antennas</u>, Vols. 1 and 2, London, UK, Peter Peregrinus, 1989.

40. L. Zaid, G. Kossiavas, J.-Y. Dauvignac, J. Cazajous, A. Papiernik, "Dual- Frequency and broad-band antennas with stacked quarter wavelength elements," IEEE Transactions on Antennas and Propagation, Volume: 47, Issue: 4, April 1999, pp. 654-660.

41. R. L. Li, G. DeJean, M. M. Tentzeris, J. Laskar, "Development and analysis of a folded shorted-patch antenna with reduced size," IEEE Transactions on Antennas and Propagation, Volume: 52, Issue: 4, Feb. 2004, pp. 555-562.

42. P. Kildal, "Artifically soft and hard surfaces in electromagnetics," IEEE Transactions on Antennas and Propagation, Volume: 38, Issue: 10, Oct. 1990, pp. 1537-1544.

43. D. C. Montgomery, <u>Design and Analysis of Experiments</u>, New York, John Wiley & Sons, Inc., 1997.

44. C. Robert, G. Casella, <u>Monte Carlo Statistical Methods</u>, New York, Springer Science + Business Media, Inc., 2004.

45. A. Taflove, S. C. Hagness, <u>Computational Electrodynamics</u>, Boston, Artech House, Inc., 2005.

46. A. Peterson, S. Ray, R. Mittra, <u>Computational Methods for Electromagnetics</u>, New York, John Wiley & Sons, Inc. – IEEE Press, 1998.

47. J. L. Volakis, A. Chatterjee, L. C. Kempel, <u>Finite-Element Method for Electromagnetics</u>, New York, John Wiley & Sons, Inc. – IEEE Press, 1998.

48. D. de Cogan, <u>Transmission Line Matrix (TLM) Techniques for Diffusion Applications</u>, London, CRC Press, 1998.

49. http://www.ansoft.com/products/hf/hfss/. October 2005.

50. http://www.sonnetusa.com/products/em/. October 2005.

51. http://www.flomerics.com/MicroStripes/. November 2005.

52. D. Pozar, <u>Microwave Engineering</u>, New York, John Wiley & Sons, Inc., 2005.

53. T. Mangold, P. Russer, "Full-wave modeling and automatic equivalent-circuit generation of millimeter-wave planar and multilayer structures," IEEE Transactions on Microwave Theory and Techniques, Volume: 47, Issue: 6, June 1999, pp. 851-858.

54. R. Araneo, S. Celozzi, "A general procedure for the extraction of lumped equivalent circuits from full-wave electromagnetic simulations of interconnect discontinuities," Proc. 15th Int. Zurich Symp., Zurich, Switzerland, Feb. 2003, pp. 419-424.

55. R. Araneo, "Extraction of broad-band passive lumped equivalent circuits of microwave discontinuities," IEEE Transactions on Microwave Theory and Techniques, Volume: 54, Issue: 1, Jan. 2006, pp. 393-401.

56. S. Jie, W. Yong, Z. Xuesheng, "A New Method of Modeling Linear Dipole Antennas for UWB Applications," 2nd International Conference on Mobile Technology, Applications, and Systems, Nov. 2005, pp. 1-4.

57. G. DeJean, M. M. Tentzeris, "Modeling and optimization of circularly-polarized patch antennas using the lumped element equivalent circuit approach," IEEE Antennas and Propagation Society International Symposium, Volume: 4, July 2004, pp. 4432-4435.

58. B. Gustavsen, A. Semlyen, "Rational approximation of frequency domain responses by vector fitting," IEEE Transactions on Power Delivery, Volume: 14, Issue: 3, July 1999, pp. 1052-1061.

59. B. Gustavsen, A. Semlyen, "Enforcing passivity for admittance matrices approximated by rational functions," IEEE Transactions on Power Delivery, Volume: 16, Issue: 1, Feb. 2001, pp. 97-104.

60. P. Russer, M. Righi, C. Eswarappa, W. J. R. Hoefer, "Lumped element equivalent circuit parameter extraction of distributed microwave circuits via TLM simulation," IEEE Microwave Theory and Techniques Society International Microwave Symposium, Volume: 2, May 1994, pp. 887-890.

61. L. Weinberg, Network Analysis and Synthesis, New York, McGraw-Hill Book Company, Inc., 1962.

62. R. B. Waterhouse, N. V. Shuley, "Dual frequency microstrip rectangular patches," IEEE Electronic Letters, Volume: 28, Issue: 7, Mar. 1992, pp. 606-607.

63. C. Degen, W. Keusgen, "Performance evaluationof MIMO systems using dual-polarized antennas," 10th International Conference on Telecommunications, Volume: 2, Feb. 2003, pp. 1520-1525.

64. B. G. Porter, L. L. Rauth, J. R. Mura, S. S. Gearhart, "Dual-polarized slot-coupled patch antennas on Duroid with teflon lenses for 76.5-GHz automotive radar systems," IEEE Transactions on Antennas and Propagation, Volume: 47, Issue: 12, Dec. 1999, pp. 1836-1842.

65. J. Huang, "The finite ground plane effect on the microstrip antenna radiation patterns," IEEE Transactions on Antennas and Propagation, Volume: 31, Issue: 4, July 1983, pp. 649-653.

66. J. Granholm, N. Skou, "Dual-frequency, dual-polarization microstrip antenna development for high-resolution, airborne SAR," Asia Pacific Microwave Conference, Dec. 2000, pp. 17-20.

67. K. Jayaraj, T. E. Noll, D. R. Singh, "RF characterization of a low cost multichip packaging technology for monolithic microwave and Millimeter Wave Integrated Circuits," URSI International Symposium on Signals, Systems, and Electronics, Oct. 1995, pp. 443-446.

68. G. Zou, H. Gronqvist, P. Starski, J. Liu, "High frequency characteristics of liquid crystal polymer for system in a package application," IEEE 8th International Symposium on Advance Packaging Materials, March, 2002, pp. 337-341.

69. G. Zou, H. Gronqvist, J. P. Starski, J. Liu, "Characterization of liquid crystal polymer for high frequency system-in-a-package applications," IEEE Transactions on Advanced Packaging, Volume: 25, Issue: 4, Nov. 2002, pp. 503-508.

70. B. Farrell, M. St. Lawrence, "The processing of liquid crystalline polymer printed circuits," IEEE Electronic Components and Technology Conference, May 2002, pp. 667-671.

71. C. Murphy, Rogers Corporation, private communication, January 2004.

72. R. Munson, "Conformal microstrip antennas and microstrip phased arrays," IEEE Transactions on Antennas and Propagation, Volume: 22, Issue: 1, Jan. 1974, pp. 74-78.

73. K. C. Gupta, R. Garg, , I. Bahl, P. Bhartia, Microstrip Lines and Slotlines, Boston, Artech House, Inc., 1996.

74. M. D. DuFault, A. K. Sharma, "A novel calibration verification procedure for millimeter-wave measurements," IEEE International Symposium on Microwave Theory and Techniques, Volume: 3, June 1996, pp. 1391-1394.

75. T. Huynh, K. F. Lee, R. Q. Lee, "Crosspolarisation characteristics of rectangular patch antennas," Electronic Letters, Volume: 24, Issue: 8, April 1988, pp. 463-464.

76. http://www.manufacturers.com.tw/telecom/Yagi-Antenna.html. September 2004.

77. http://www.starantenna.com. September 2004.

78. J. Huang, "Planar microstrip Yagi array antenna," IEEE Antennas and Propagation Society International Symposium, Volume: 2, June 1989, pp. 894-897.

79. A. Densmore, J. Huang, "Microstrip Yagi antenna for mobile satellite service," IEEE Antennas and Propagation Society International Symposium, Volume: 2, June 1991, pp. 616-619.

80. J. Huang, A. Densmore, "Microstrip Yagi antenna for mobile satellite vehicle application," IEEE Transactions on Antennas and Propagation, Volume: 39, Issue: 7, July 1991, pp. 1024-1030.

81. S. Padhi, M. Bialkowski, "Investigations of an aperture coupled microstrip Yagi antenna using PBG structure," IEEE Antennas and Propagation Society International Symposium, Volume: 3, June 2002, pp. 752-755.

82. P. R. Grajek, B. Schoenlinner, G. M. Rebeiz, "A 24 GHz high-gain Yagi-Uda antenna array," IEEE Transactions on Antennas and Propagation, Volume: 52, Issue: 5, May 2004, pp. 1257-1261.

83. D. Gray, J. Lu, D. Thiel, "Electronically steerable Yagi-Uda microstrip patch antenna array," IEEE Transactions on Antennas and Propagation, Volume: 46, Issue: 5, May 1998, pp. 605-608.

84. S. Ke, K. Wong, "Rigorous analysis of rectangular microstrip antennas with parasitic patches," IEEE Antennas and Propagation Society International Symposium, Volume: 2, June 1995, pp. 968-971.

85. G. R. DeJean, M. M. Tentzeris, "A new high-gain microstrip Yagi array antenna with a high front-to-back (F/B) ratio for WLAN and millimeter-wave applications," accepted for publication in the IEEE Transactions on Antennas and Propagation.

86. K. Barry, "Counterfeits and Counterfeiters: The Ancient World," Available on-line at: http://www.ancient-times.com/newsletters/n13/n13.html. July 2003.

87. D. W. Bauder, Personal Communication.

88. P. P. Ewald, Ann. der Physik, Volume: 49, 1915, pp. 1-56.

89. C. W. Oseen, "Uber die Wechrelwirkung zwischen zwei elektrischen Dipolen und uber die Drehung der Polarisationsebene in Kristallen und Flussigkeiten," Ann. der Physik, Volume: 48, 1915, pp. 1-56.

90. E. Wolf. "A generalized extinction theorem and its role in scattering theory," Coherence and Quantum Optics edited by L. Mandel and E. Wolf (eds.), New York, Plenum, 1973.

91. L. Tsang, J.A. Kong, and K.H. Ding. <u>Scattering of Electromagnetic Waves</u>, New York, Wiley Interscience, 2000 & 2001.

92. L. Tsang, J. A. Kong, R. Shin. <u>Theory of Microwave Remote Sensing</u>, New York, Wiley Interscience, 1985.

93. M. Born, E. Wolf, <u>Principles of Optics: Electromagnetic Theory of Propagation, Interference and Diffraction of Light</u>, Oxford, Pergamon Press, 1975.

94. M. Nieto-Vesperinas, <u>Scattering and Diffraction in Physical Optics</u>, New York, John Wiley & Sons, Inc., 1991.

95. S. K. Cho, <u>Electromagnetic scattering</u>, New York, Springer-Verlag, 1990.

96. CST Corp. Microwave Studio. Available on-line at: http://www.cst.de/Content/Products/MWS/Solvers.aspx. August 2005.

97. P. Xu, L. Tsang. "Scattering by rough surface using a hybrid technique combining the multilevel UV method with the sparse matrix canonical grid method," Radio Science, Volume: 40, 2005.

98. W. C. Chew. <u>Waves and Fields in Inhomogenous Media</u>, New York, John Wiley & Sons, Inc. – IEEE Press, 1999.

99. Microwave Engineering Europe. CAD benchmark. October 2000 – February 2001. Available on-line at: http://i.cmpnet.com/edtn/europe/mwee/pdf/CAD.pdf. August 2005.

100. A.J. Menezes, P. C. van Oorschot, S. A. Vanstone, <u>Handbook of Applied Cryptography</u>, CRC Press, 1996.

101. R.L. Rivest, A, Shamir, L. Adleman. "A method for obtaining digital signatures and publickey cryptosystems," Communications of the Association for Computing Machinery (ACM), Volume: 21, Issue: 2, 1998, pp. 120-126.

102. N. Koblitz, "Elliptic Curve Cryptosystems," Mathematics of Computation, Volume: 48, Issue: 177, 1987, pp. 203-209.

103. D. Kirovski, "Toward an Automated Verification of Certificates of Authenticity," Association for Computing Machinery (ACM) Electronic Commerce, 2004, pp.160-169.

104. G. DeJean et.al. "Making RFIDs Unique – Radio Frequency Certificates of Authenticity," presented at the 2006 IEEE Antennas and Propagation Society International Symposium in Albuquerque, NM.